Grant Allen

Flashlights on Nature

Grant Allen

Flashlights on Nature

ISBN/EAN: 9783337026967

Printed in Europe, USA, Canada, Australia, Japan

Cover: Foto ©berggeist007 / pixelio.de

More available books at **www.hansebooks.com**

FLASHLIGHTS ON NATURE

Flashlights on Nature

By

Grant Allen

Author of " The Story of the Plants," etc.

With 150 Illustrations by

Frederick Enock

London

George Newnes, Limited

Southampton Street, Strand

1899

CONTENTS AND ILLUSTRATIONS

FLASHLIGHTS ON NATURE

I

THE COWS THAT ANTS MILK

DON'T let my title startle you; it was Linnæus himself who first invented it. Everybody knows the common little "green-flies" or "plant-lice" that cluster thick on the shoots of roses; and most people know that these troublesome small insects (from the human point of view) are the true source of that shining sweet juice, rather slimy and clammy, that covers so many leaves in warm summer weather, and is commonly called honey-dew. A good many people have heard, too, that ants use the tiny green creatures in place of cows, coaxing them with their feelers so as to make them yield up the sweet and nutritious juice which is the ants' substitute for butter at breakfast. But comparatively few are aware how strange and eventful is the brief life-history of these insignificant little beasts which we destroy by the thousand in our flower-gardens or conservatories with a sprinkle of tobacco-water.

A

To the world at large, the aphides, as we call them,
are mere nameless nuisances — pests that infest
our choicest plants ; to the eye of the naturalist,
they are a marvellous and deeply interesting
group of animals, with one of the oddest pedi-
grees, one of the queerest biographies, known to
science.

I propose, therefore, in this paper briefly to
recount their story from the cradle to the grave ;
or, rather, to be literally accurate, from the time
when they first emerge from the egg to the moment
when they are eaten alive (with some hundreds of
their kind) by one or other of their watchful enemies.   In this task I shall be aided not a little
by the clever and vivid dramatic sketches of the
Aphides at Home, which have been prepared for
me by my able and watchful collaborator, Mr.
Frederick Enock, an enthusiastic and observant
naturalist, who thinks nothing of sitting up all
night, if so he may catch a beetle's egg at the
moment of hatching ; and who will keep his eye
to the microscope for twelve hours at a stretch,
relieved only by occasional light refreshment in the
shape of a sandwich, if so he may intercept some
rare chrysalis at its moment of bursting, or behold
some special grub spin the silken cocoon within
whose case it is to develop into the perfect winged
insect.

Rose-aphides, or " green-flies," as most people
call them, are, to the casual eye, a mere mass of
living " blight "—a confused group of tiny trans-

lucent insects, moored by their beaks or sucking-
tubes to the shoots of the plant on which they have
been born, and which they seldom quit unless
forcibly ejected.   For they are no Columbuses.
The spray of rose-bush figured in sketch No. 1
shows a small part of one such numerous household in quiet possession of its family tree, and

NO. 1.—A BRANCH OF THE FAMILY TREE.

engaged, as is its wont, in sucking for dear life at
the juices of its own peculiar food-plant.   You
will observe that they are clustered closest at the
growing-point.   Each little beast of this complex
family is coloured protectively green, so as to be
as inconspicuous as possible to the keen eyes of its
numerous enemies ; and each sticks to its chosen
twig with beak and sucker as long as there is any-

thing left to drink in it, only moving away on its six sprawling legs when its native spot has been drained dry of all nutriment.

We often talk metaphorically of vegetating : the aphis vegetates. Indeed, aphides are as sluggish in their habits and manners as it is possible for a living and locomotive animal to be : they do not actually fasten for life to one point, like oysters or barnacles ; but they are born on a soft shoot of some particular plant ; they stick their sucking-tube into it as soon as they emerge ; they anchor themselves on the spot for an indefinite period ; and they only move on to a new "claim" when sheer want of food or *force majeure* compels them. The winged members are an exception : *they* are founders of new colonies, and are now on their way to some undiscovered Tasmania.

And, indeed, as we shall see, these stick-in-the-mud creatures have yet, in the lump, a most eventful history—a history fraught with strange loves, with hairbreadth escapes, with remorseless foes, with almost incredible episodes. They have enemies enough to satisfy Mr. Rider Haggard or the British schoolboy. If you look at No. 2, you will see the first stage in the Seven Ages of a rose-aphis family. The cycle of their life begins in autumn, with the annual laying of the winter eggs ; these eggs are carefully deposited on the leaf-buds of some rose-bush, by a perfect wingless female, at the first approach of the cold weather. I say a perfect wingless female, because, as I shall explain hereafter, most aphides (and especially all the summer crops or generations that appear with such miraculous rapidity on our roses and fruit-trees) are poor fatherless creatures ; waifs and strays, budded out vegetatively like the shoots of a plant.

About this strange retrogressive mode of reproduction, however, I shall have more to tell you in due time by-and-by ; for the present, we will confine ourselves to the immediate history of the autumn brood, which is regularly produced in the legitimate fashion, as the result of an ordinary insect marriage between perfectly developed males and females. As October approaches, a special generation of such perfect males and females is produced by the un-

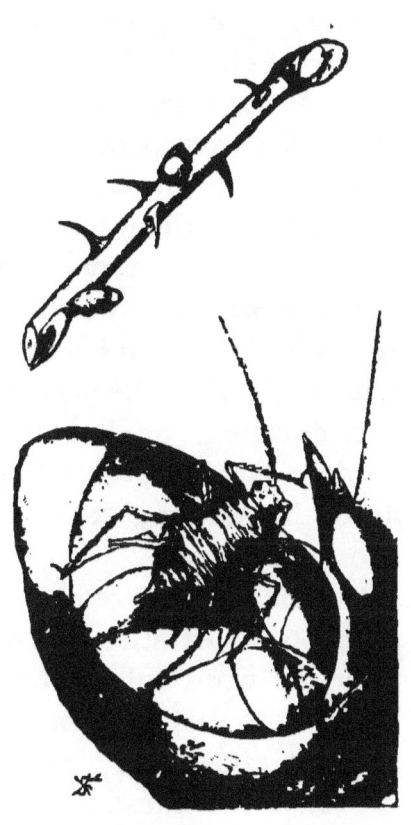

NO. 2.—WORN-OUT MOTHER— LAYING HER LAST EGG.

wedded summer green-flies ; and the females of this brood, specially told off for the purpose, lay the winter eggs, which are destined to carry on the life of the species across the colder months, when no fresh shoots for food and drink are to be found in the frozen fields or gardens.

The eggs, so to speak, must be regarded as a kind of deferred brood, to bridge over the chilly time when living aphides cannot obtain a livelihood in the open. In No. 2 we see, above, a rose-twig with its leaf-buds, which are undeveloped leaves, inclosed in warm coverings, and similarly intended to bridge over the winter on behalf of the rose-bush. On this twig, then, we have the winter eggs of the aphis, mere dots represented in their natural size; they are providently laid on the bud, which in early spring will grow out into a shoot, and thus supply food at once for the young green-flies as they hatch and develop. So beautifully does Nature in her wisdom take care that blight in due season shall never be wanting to our Marshal Niels and our Gloires de Dijon !

In the same sketch, too, we have, below, a pathetic illustration, greatly magnified, of the poor old worn-out mother, a martyr to maternity, laying her last egg in the crannies of the bud she has chosen. I say "a martyr to maternity" in solemn earnest. You will observe that she is a shrivelled and haggard specimen of over-burdened motherhood. The duties of her station have clearly been too much for her. The reason is that she literally uses herself up in the production of offspring; which is not surprising, if you consider the relative size of egg and egg-layer. When this model mother began to lay, I can assure you she was fat and well-favoured, as attractive a young green-fly as you would be likely to come across in a day's march on the surface of a rose-twig. But once she sets to work, she lays big eggs with a will (big, that is to say, compared with her own size), till she has used up all her soft internal material ; and when she has finished, she dies—or, rather, she ceases to be ; for there is nothing left of her but a dried and shrivelled skin.

During the winter, indeed—in cold climates at least—the race of aphides dies out altogether for the time being, or only protracts an artificial existence in the heated air of green-houses and drawing-rooms. The species is represented at such dormant periods by the fertilised eggs alone, which lie snug among the folds or scales of the buds till March or April comes back again to wake them. Then, with the first genial weather, the eggs hatch out, and a joyous new brood of aphides emerges. And here comes in one of the greatest wonders ; for these summer broods do not consist, like their parents in autumn, of males and females, but of imperfect mothers—all mothers alike, all brother-less sisters, and all budding out young as fast as they can go, without the trouble and expense of a father. They put forth their progeny as a tree puts forth leaves, by mere division. The new broods thus produced are budded out tail first, as shown in No. 3, so that all the members of the family stand with their heads in the same direction, the mother moving on as her offspring increases ;

and since each new aphis instantly begins to fix
its proboscis into the soft leaf-tissue, and in turn
to bud out other broods of its own, you need not
wonder that your favourite roses are so quickly
covered with a close layer of blight in genial
weather.

To say the truth, the rate of increase in aphides
is so incredibly rapid, that one dare hardly mention
it without seeming to exaggerate. A single industrious little green-fly, which devotes itself with
a quiet mind to eating and reproduction, may easily
within its own lifetime become the ancestor of some
billions of great-grandchildren. It is not difficult
to see why this should be so. The original parent
buds out little ones from its own substance at a
prodigious rate; and each of these juniors, reaching maturity at a bound, begins at once to bud out
others in turn, so that as long as food and fine
weather remain the population increases in an
almost unthinkable ratio. Of course, it is the extreme abundance of food and the ease of living
that result in this extraordinary rate of fertility;
the race has no Malthus to keep it in check—each
aphis need only plunge its beak into the rose-shoots
or leaves and suck; it can get enough food without
the slightest trouble to maintain itself and a numerous progeny. It does not move about recklessly,
or use up material in any excessive intellectual
effort; all it eats goes at once to the production
of more and more aphides in rapid succession.

Many things, however, conspire to show that

aphides did not always lead so slothful a life : they are creatures with a past, the unworthy descendants of higher insects, which have de-generated to this level through the excessive abundance of their food, and through their adoption of what is practically a parasitic habit. When life is too easy, men and insects invariably degenerate : struggle is good for us. One of these little indications of a higher past Mr. Enock has given us in the upper part of sketch No. 3. For some members of the brood go through regular stages of grub and chrysalis, like any other flies ; or, if you wish to be accurately scientific, pass through the usual forms of larva and pupa, before they reach the full adult con-

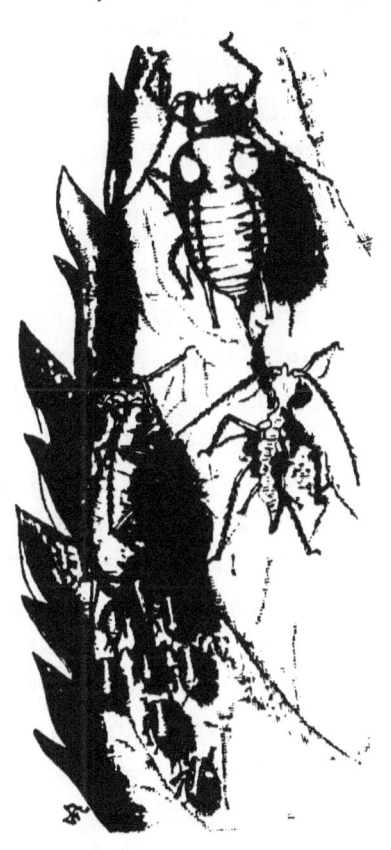

NO. 3.—BUDDING MOTHER—PRODUCING A FATHERLESS BROOD.

dition. This, of course, shows them to be the descendants of higher insects which underwent the common metamorphosis of their kind. But most of the budded-out, fatherless broods in summer are

produced ready-made, without the necessity for
passing through larval or infantile stages. Or
rather, they never grow up : they merely moult ;
and they produce more young while they are still
larvæ. They are born fully formed, and proceed
forthwith to moor themselves, to feed, and to bud
out fresh generations, without sensible interval. In
No. 3 we have various stages in the development
of the spring brood. Above we see the pupa, or
chrysalis, produced from a grub (not very grub-like
in shape), which has sprung from an egg; and on
the right, below, we see the shrivelled larval skin
from which it has just freed itself. This particular
aphis was thus born as a six-legged larva from an
autumn egg; it passes through the intermediate
form of a pupa, or chrysalis; and it will finally
develop into a winged "viviparous" female, such
as you see in No. 4 below, putting out its young
alive as fast as ever its wee body can bud them.
You may observe, however, that in the case of
aphides there is no great difference of form between
the three successive stages. Larva, pupa, and fly
are almost identical.

In No. 4, again, we have a portrait from life of
such a *winged* female, the mother of a numerous
fatherless progeny ; for both winged and wingless
forms are produced through the summer. She is
round and well-fed, as becomes a matron. Observe
in particular the curious pair of tubes on the last
few rings of her back; these are the organs for
secreting nectar or *honey-dew*, a point about which

I shall have a good deal more to say presently. A winged female like this may fly away to another rose-bush to become the foundress of a distant colony. The same illustration also shows, in a greatly enlarged form, her beak or sucking apparatus, which con-
sists of four sharp lance-like siphons, enclosed in a protective sheath or proboscis, and admirably adapted both for piercing the rose-twig and for draining the juices of your choicest crimson ramblers. The aphis sticks in the point as if it were a needle, and then sucks away vigorously at the rose-tree's life-blood. You can watch her so any day with a common small mag-

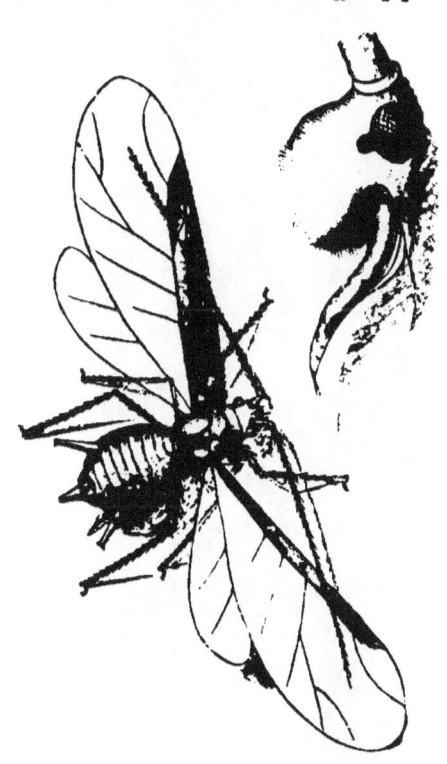

NO. 4.—WINGED FEMALE—THE FOUNDRESS OF A COLONY.

nifier, and see how, like the lady at Mr. Stiggins' tea meeting, she "swells wisibly" in the process. Indeed, aphides are always beautiful objects for the microscope or pocket lens, with their pale, transparent green bodies, their bright black eyes, their jointed hairy legs, their delicate feelers, and their marvellous honey-tubes; and it will not be my fault if you still continue to regard them as nothing more than the "nasty blight" that destroys your roses.

NO. 5. UNNATURAL LODGER EATS HIS HOSTESS OUT OF HER SKIN.

Do not for a moment suppose, however, that you and your gardener, with his spray and his tobacco-water, are the only enemies the rose-aphis possesses. The name of her foes is legion. She is devoured alive, from without and from within, by a ceaseless horde of aggressive belligerents. The most destructive of these enemies are no doubt the lady-birds, which, both in their larval and their winged forms, live almost entirely on various kinds of green-fly. This practical fact in natural history is well known to hop-growers, for the dreaded "fly" on hops is an aphis; its abundance or otherwise governs the hop market, and Kentish farmers are keenly aware that a certain particular lady-bird eats the "fly" by millions, on which account they protect and foster the lady-bird, thus leaving the two insects, the parasite and the carnivore, to fight it out in their own way between them.

But No. 5 introduces us to a still more insidious though less dangerous foe : an internal parasite which lays its eggs inside the body of the bud-producing female. There the grub hatches out, and proceeds to eat up its unwilling hostess, alive, *from within.* In the sketch, we have an illustration, below, of an aphis which has thus been compelled to take in a stranger to board and lodge in her stomach ; while the top figure shows how the lodger, after eating his hostess out, eats himself out into the open air through her empty skin. If you look out closely for such haunted green-flies, in-habited by a parasite—most often an ichneumon fly—you will find them in abundance on the twigs of rose-bushes. They have a peculiar swollen, quiescent look, and a brownish colour.

No. 6 shows us another such fierce enemy at work. This formidable insect tiger is the larva of the wasp-fly ; he is a savage carnivore, who moors himself by his tail end, stretches out to his full length, and swoops down upon his unsuspecting prey from above ; and being blessed with a good appetite, he can get rid of no fewer than 120 aphides in an hour. As he probably eats all day, with little intermission for rest and digestion, this gives a grand total of about 1500 or 1600 victims at a sitting. However, the remaining aphides go on budding away as fast as ever to make up the deficiency, so the loss to the race is by no means irreparable. " *Il n'y a pas d'homme nécessaire,*" Napoleon used to say; and the principle is even more true as applied to the green-flies. If a few millions die, their place is soon filled again.

NO. 6.—TRAGIC ENEMY WHO DEVOURS 120 PER HOUR.

Look once more at No. 6, and you will see that while the tiger-like enemy is engaged in hoisting and devouring one unfortunate aphis, its neighbour below, heedless of the tragedy, is quietly engaged in blowing off honey-dew.

This blowing-off of honey-dew leads me on direct to the very heart of my subject; for it is as manufacturers of honey-dew and as cows to the ants that aphides base their chief claim to attention. If they did not produce this Turkish delight of the insect world, nobody would have troubled to study them so closely. Let us go on to see, then, what is the origin and meaning of this curious and almost unique secretion.

If you examine the leaves of a lime-tree or a rose-bush in warm summer weather you will find them covered all over with a soft sticky substance, sweet to the taste, and spread in a thin layer upon the surface of the foliage. This sweet stuff is honey-dew, and it is manufactured solely by various kinds of aphides, without whose trade-mark none other is genuine. Why do they make it ? Not, you may be sure, out of pure unselfish moral desire to benefit the ants and other beasts that like it. In the animal world, nothing for nothing is the principle of con-duct. The true secret of the origin of honey-dew appears to be this. Aphides live entirely off a light diet of vegetable juices; now, these juices are rich in compounds of hydrogen and carbon, especially sugar (or rather, to be strictly scientific, glucose), but are relatively deficient in nitrogenous materials, which last are needed as producers of movement by all animals, however sluggish. In order, therefore, to procure enough nitrogenous matter for its simple needs, your aphis is obliged to eat its way through a quite superfluous amount of sweets, or of sugar-forming substances. It is almost as though we ourselves had to swallow daily a barrel of treacle so as to reach at the bottom an ounce of beefsteak. To get rid of this surplus of sugar (or rather, un-digested glucose) almost all aphides (for they are a large family, with many separate kinds) have acquired a pair of peculiar organs, known as honey-tubes, on the backs of their bodies. Sometimes, when distended with superfluous food, they simply blow out the honey-dew secreted by these tubes on to the leaves below them.

The aphis in No. 6 is represented at the moment when it is thus ridding itself of its excessive sweetness. But honey-dew is sticky, and apt to get in the way; it may clog one's legs, or interfere with one's proboscis: so the aphides prefer as a rule to retain it prudently till some friendly animal, with a taste for sweets, steps in to relieve them of the unpleasant tension. The animal which especially performs this kind office for the rose-aphis is the

NO. 7.—AN ANT MILKING A ROSE-APHIS OF ITS HONEY-DEW.

garden ant; and No. 7 represents such an ant in the very act of tapping and caressing an aphis with its feelers, in order to make her yield up on demand her store of honey. The process is ordinarily described as "milking."

You must understand, of course, that neither aphis nor ant is actuated by purely philanthropic considerations; this is a case of mutual accommo-

dation. The aphis wants to get rid of a troublesome waste product which is apt to clog it. The
ant wants to secure that waste product as a valuable
food-stuff. Hence, from all time, an offensive and
defensive alliance of the profoundest type has been
mutually struck up between ants and aphides.
How far this alliance has gone is truly wonderful.
The ants not merely "milk" the aphides, but actually collect them together in herds and keep them
in parks as domestic animals. Nay, more; as Sir
John Lubbock has pointed out, different kinds of
ants domesticate different breeds of aphides, as each
is suited to the other's conditions. The common
black garden ant attends chiefly to the aphides
which frequent twigs and leaves, such as this very
rose-aphis—for the black ant is a rover and a good
tree-climber; he is much given to exploring ex-
peditions over the surface of plants in search of
honey, and he is not particular whether he happens
to gather it from flowers or from insects. The
brown ant, on the other hand, goes in rather for
such species of aphides as frequent the crannies
in the bark of trees; while the little yellow ant,
an almost subterranean race, living underground
among the grass roots in meadows, "keeps flocks
and herds" (says Lubbock) "of the root-feeding
aphides." All these facts you can verify for yourself with very little trouble.

It is most interesting to watch a black ant on the
prowl after honey-dew. He is evidently led on to
the herd by smell, for he mounts the stem where

B

the aphides live in a business-like way, and goes
straight to the point, as if he knew what he was
after.   When he finds an aphis that looks likely,
he strokes and caresses her gently with his antennæ
(as you see in the sketch), coaxing her to yield up
the coveted nectar.   The aphis, on her side, glad
to receive his polite attentions, and accustomed to
the signal, exudes a clear drop of her surplus sweet,
which the ant licks up with its jaws greedily.   But
ants do much more than this in the way of aiding
and protecting their "cows."   They really appropriate them.   Often they build, with mud, covered
ways or galleries up to their particular herds, and
erect earthen cowsheds above them; they also
fight in defence of their flocks, as a Zulu will fight
for his oxen, or an Arab for his camels.   Their
foresight is almost human: for when the winter
eggs are laid, the ants will transport them into
their nest, to keep them safe against frost; and
when summer comes again, they will carry them
out with care, and place them in the sun to hatch
on the proper food-plant.   Could man himself
show greater prudence and forethought than these
mites of herdsmen ?

"The eggs," says Sir John Lubbock, "are laid
early in October on the food-plant of the insect.
They are of no direct use to the ants; yet they
are not left where they are laid, exposed to the
severity of the weather and to innumerable dangers,
but brought into the nests, and tended with the
utmost care through the long winter months till

the following March," when they are brought out again and placed on their special food-plant.

Lubbock even notes that ants have domesticated a far larger variety of other animals than we ourselves have. Our list includes at best the horse, the dog, the cat, the cow, the camel, the sheep, the llama, the alpaca, the goat, the hen, the duck, the goose, the bee, the silkworm, and a dozen or so others ; while ants have domesticated no fewer than 584 different kinds of crustaceans and insects, including beetles, flies, and mites, some of which have lived for so many generations in the dark galleries of the ant-hills that they have become totally blind, as happens almost always, in the long run, with underground animals.

During the live-long summer the aphides go on, eating and drinking, budding out new broods with inexhaustible fertility. They settle down calmly on the spot where they were born, they stick to it for life, and they seldom move away from their native twig unless somebody pushes them, for though they have legs, they do not care to use them except on extreme provocation. But when autumn arrives "a strange thing happens." Broods of perfect winged males and wingless females are then produced ; and the males of these, like almost all other insects, take a marriage flight, find their predestined mates, and become with them the parents of the dormant eggs which outlive the year, and carry on the race to the succeeding summer. While warm weather lasts, few or no males are budded out; it is only when the cold threatens to destroy the entire colony that little husbands are born, so as to give rise to eggs which may bridge over the gulf between summer and summer. If you keep the insects warm, however, and supply them with abundant food (as in a conservatory), they will go on producing imperfect females and fatherless broods, without intermission, for many years together. The egg-laying generation is thus shown to be merely a device for meeting the adverse chances of winter; the budding process suffices well enough, as long as warmth and food render the possibility of freezing or starvation unimportant.

On the other hand, the eggs and the brood born from them revert to the earlier habit of the race, when it was still an active, free-flying type, before it had been demoralised by acquiring its sedentary, parasitic habits. They hatch out into active little six-footed or six-legged larvæ, which again, in some cases, give rise to very similar chrysalis forms, and finally develop into the "viviparous" or budding females. Whenever a species earns its livelihood with too little exertion, it invariably degenerates, and often grows small, unintelligent, and vastly prolific; for superior races have relatively small families, while inferior races reproduce by the million. The mites which infest cheese and other food-stuffs are an exactly analogous case to that of the aphides, for they are degenerate spiders, grown small and prolific through the excessive ease of life afforded them by always settling in a cheese, all ready-made food for them, without the trouble or exertion of hunting.

Creatures which reproduce at such a rate, however, invariably pay the penalty for their rapid increase by an equally rapid and enormous death-rate; were it otherwise, the offspring of a single pair of codfish (with their million eggs) would soon turn the sea into one solid mass of cod; while the

NO. 8.—COMIC ENEMY WHO POSES AS OLD-CLOTHES MAN.

descendants of a single viviparous aphis would cover the earth with a ten feet thick layer of teeming green-flies. However, Nature has remedies in store for them. Storms of rain and hail kill myriads of aphides; sudden changes of weather wilt them and nip them up; innumerable enemies make an honest livelihood out of them. Another of these ubiquitous foes is graphically represented in No. 8—the grub of the lace-wing fly, a sort of insect old-clothes man, which covers its back with the cast-off skins of its discarded victims. This is a clever device to enable it to escape observation. The larva, which is a fat and juicy morsel, catches aphides wholesale, and sucks their life-blood; when he has drained them dry, he hoists up their skins on to his back with his jaws, by way of overcoat. Then the hooks or spines on his back (shown above) hold them in place for a time, while the larva bends over and spins a few threads of web across them, to weave them into a neat and compact garment. Thus securely clad, he is hidden from view: he looks much like a twig covered with aphides, and avoids to some extent the too pressing attentions of his own enemies. Observe in this sketch the characteristic unconcern of the aphis who is destined to be his next victim.

Birds also destroy large numbers of aphides. You can see them picking them off in the bean-fields in summer.

It is lucky for us that these insect pests have so abundant a supply of natural enemies; for man, by himself, is almost powerless against them. Strange to say, and paradoxical as it sounds, it is the smallest enemies that we always find most difficult to extirpate. Lions and tigers we can kill off without difficulty; they can be shot and exterminated. Wolves and hyenas give us a little more trouble; while against rabbits, our resources are taxed to the utmost. A plague of rats and mice, or of tiny field-voles, can hardly be combated with any hope of success; while locusts and Colorado beetles devastate our crops with practical impunity.

When it comes to aphides, we are quite unable to cope with the infinite numbers of our infinitesimal foes; and if we take the microscopic creatures which cause cholera, typhoid fever, and other zymotic diseases, we may keep out of their way, it is true, or may isolate the objects in which they breed and store their germs, but we are practically without means to kill or hurt them. The larger the foe, the more easily is he met; the smaller our enemy, the more difficult is he to extirpate. We killed off the American buffalo (or bison) in a single generation; a thousand years would probably fail to kill off the insignificant little aphides that infest our roses.

In the case of one member of the family at least the experiment has been tried on a gigantic scale in France, and as yet with comparatively small results. For the dreaded phylloxera which attacks the vines is, in fact, an aphis; and though immense rewards have been offered by the French Assembly for any good remedy against phylloxera, the only successful plan as yet proposed has been that of planting healthier and sturdier American vines, which resist the little beast a good deal better than the effete and worn-out European species. But many other members of the family wage war with distinguished success against the British farmer. The little black "colliers" which attack our bean crops are a species of aphis; so are the "blight" of apple-trees, the "fly" on turnips, and the most familiar parasites of the hop, the cabbage, the pear, and the potato. It is well for us, therefore, that the aphides have roused against them so many natural enemies among the birds and insects, or our crops would be destroyed by their persistent efforts. The ichneumon-flies alone kill their millions yearly; and the lady-birds well deserve their popular esteem for the good they do in keeping down the ever-increasing numbers of these voracious insects.

Yet, mischievous as they are, the tiny green aphides are well deserving of study, both for their personal beauty and their singular life-history. Everybody can observe them, because they are practically everywhere. If you have a garden, they swarm on every bush. If you grow flowers in your window, they live in every pot. If you content yourself with an occasional bunch of roses or geraniums, you will find them, if you look, sucking away contentedly on the leaves of the rosebuds. Even in London parks or squares you may watch the industrious ants creeping slowly up the stems to milk their wee green cows; you may see with the naked eye, or still better with a pocket lens, the grateful aphis exude a tiny drop of limpid honey from its translucent tubes, and the ant lick it up with unmistakable gusto. Go out into the parks or gardens and examine it for yourself; for every one of the facts I have mentioned in this paper can be verified with ease, if only you have patience.

II

A PLANT THAT MELTS ICE

IF you have ever visited the Alps in early
spring, you will know well by sight the dainty
little ·nodding bells of the alpine soldanella
—twin flowers on one stalk, like fairy tocsins,
which push their heads boldly through the ice
of the *névé*, and form a border of blue blossoms
on the edge of the snow-sheet. Most people, to
be sure, visit the Alps in August ; and they go too
late. Autumn is the time when heather purples
our bleak northern moors, but when the central
mountain chain of Europe, so glorious in April,
has become comparatively green and flowerless.
If you wish to see what nature can do in the way
of rock-gardens, however, you should go to Switzer-
land in early spring. It is then that blue gentians
spread vast girdles of blossom over the alpine pastures ; then that the green slopes on the mountain
sides are yellowed by globe-flowers ; then that the
poet's narcissus stars with its white petals and
scents with its sweet perfume the rich meadows
on the spurs of the lesser ranges. Higher up,
sheets of creeping rock-plants, close clinging to
the uneven surface, fall in great cataracts of pink

and blue over the steep declivities. As the snow
melts, upward, the flowers open in zones, one after
another, upon the mountain sides, so that you can
mark your ascent by the variations in the flora,
and the different successive stages of development
reached by the most persistent kinds at various
levels.

There is one adventurous little plant, however,
among these competing kinds, which in its eagerness to make the most of the short alpine summer
does not even wait, like its neighbours, for the
melting of the snow, but, vastly daring, begins to
grow under the surface of the ice-sheet, and melts
a way up for itself by internal heat, like a vegetable
furnace. It may fairly be called a slow-combustion
stove, not figuratively, but literally. It burns itself
up in order to melt the ice above it. This wonderful plant is the alpine soldanella, the hardest
and one of the prettiest of mountain flowers ; it
opens its fringed and pensile blue blossoms in the
very midst of the snow, often showing its slender
head above a thin layer of ice, where it fear-
lessly displays its two sister bells among the frozen
sheet which still surrounds its stem in the most
incredible fashion.

So much every tourist to the Alps in May
must have noticed for himself, for whenever he
reaches the edge of the melting ice-sheet he can
see the ice pierced by innumerable twin pairs
of these dainty and seemingly delicate blossoms.
Comparatively few observers, however, have proceeded to notice that the soldanella, fragile as it

is, actually forces itself up through a solid coat of ice, not exactly by hewing its way, but by melting a path for itself in the crystal sheet above it. Yet such is really the case ; it warms the ice as it goes. The buds begin to grow on the frozen soil before the ground is bare, under the hardened and compressed snow of the *névé*— which at its edge is always ice-like in texture. They then bore their way up by internal heat (like that of an animal) through the sheet that covers them ; and they often expand their delicate blue or white blossoms, with the scalloped edges, in a cup-shaped hollow above, while a sheet of refrozen ice, through which they have warmed a tunnel or canal for themselves, still surrounds their stems and hides their roots and their flattened foliage. This is so strange a miracle of nature that it demands some explanation ; the method by which the soldanella obtains its results is no less marvellous than the results themselves which it produces.

The winter leaves of soldanella, which hibernate under the snow just as truly as the squirrel or the dormouse hibernates in its nest, are large, leathery, tough, and evergreen. They are, in fact, just living reservoirs of fuel (like the fat of the dormant bear), which the plant lays by during the heat of summer in order to burn it up again in spring for the use of its flowers. When I use this language, you will think at first I am speaking figuratively. But I am not; I mean it in just as literal a sense as when I say that the coal in

the tender of a locomotive serves as fuel for the
engine, or that the corn in the bin of a stable
serves as fuel to heat the horse's body. These
leaves contain material laid by for burning ; and
it is by burning that material up at the proper
period that the soldanella manages to melt its
way out of the wintry ice-sheet, and so to steal
a march upon competing species.

The process requires explanation, I admit ; let
us try to understand it. Everybody knows, as
a matter of common experience, that animals are
warmer in winter than the air which surrounds
them ; warm-blooded animals, that is to say, which
form the only class most people trouble about.
Not everybody knows, however, that the same
thing is more or less true of plants as well—
that many plants have the power of evolving
heat for themselves in considerable quantities.
But this is actually true ; indeed, all growing
parts of a stem or young leaf-shoot must necessarily be slightly warmer than the air around
them.   For, when you come to think of it,
whence do animals derive their heat ? " From
the oxidation of their food," the small boy of
the day, crammed full of knowledge, will tell
you, glibly. And what do you mean by oxidation but very slow burning ? You may take a
load of hay, and set a match to it, and it will
burn at once quickly, by combining with the
oxygen of the air in the open ; or you may, if
you choose, give it to a pair of horses to eat
instead, and then it will burn up slowly, by

combining with the oxygen of the air in their
bodies. Lungs, in fact, are mere devices for taking
in fresh oxygen, which then combines with the
food or fuel in the blood of the animal.

A century ago, Count Rumford pointed out that
you might burn your hay as you chose, either
in a horse or in a steam-engine; and that in
either case you produced alike heat and motion.
What we call fuel is just carbon and hydrogen,
separated from oxygen; and what we call burning
or combustion is just the re-union of the oxygen
with the other elements, accompanied by a giving-
off of heat equivalent in amount to that originally
required in order to separate them.

Now, the foodstuffs of most animals are plants
or parts of plants, especially seeds or grains, as
well as the rich stores of starch or oil laid by in
roots, bulbs, and tubers. These are all of them
reservoirs of food or fuel, produced by the plant
for its own future growth, and meant hereafter to
sprout or germinate. All seeds, when they begin
to quicken, unite with oxygen and evolve heat;
and this heat is just the same in nature, whether it
happen to be set free within or without an animal
body. If you give an ox corn, he will oxidise it
internally and warm his own body with it; but if
you let it germinate, it will oxidise itself, and so
produce a very small but slow fire, which warms
both the corn and the space around it. Similarly,
all growing shoots combine with oxygen, and,
therefore, rise in temperature. In early spring,
when the ground just teems with sprouting seeds

and swelling buds, with growing bulbs or shooting tubers, the temperature of the soil is sensibly raised ; and this very heat, evolved by germination, becomes itself in turn a cause of more germination ; each seed and root and bulb and sucker. helps to warm and start all the others. Spring largely depends upon the warmth thus produced. The earth, during this orgy of growth, is warmer by a good deal than the air about it ; warmer even than it is in summer weather—indeed, were it not for the number of plants which thus start growing at once, growth would be almost impossible in very cold countries. Like roosting fowls, they warm one another.

You think, however, the amount of heat that can be thus evolved must be very insignificant. By no means. Take an example in point. What do we mean by malting ? We collect together a number of seeds or grains of barley, we wet them thoroughly, and allow them to begin germinating. Each grain individually gives out only a small amount of heat, it is true : but when many of them lie together, the total volume of heat produced is very great, and the amount would be even greater if it were not artificially checked at a certain stage : for the maltster does not wish his malt to be "over-heated." Malt, then, is nothing more than sprouting barley ; and the heat it begets in the process of malting shows us very clearly how much warmth exists in sprouting seeds, or in the growing portions of young plants, buds, shoots, and tubers.

At the risk of seeming tedious in this preliminary explanation, I must also add that flower-buds
and flower-stems which grow and open very rapidly
must similarly use up oxygen in their growth, and
therefore distinctly rise in temperature. In a very
few large and conspicuous flowers, such as the big
white calla lily, this rise in temperature during the
flowering period can be measured even with an
ordinary thermometer. No bud can open without
giving out heat ; and the amount of heat is sometimes considerable.

And now, I hope, we are in a position to
understand how soldanella acts, and why it does
so. It is a plant which grows under peculiarly
trying conditions. It has to eke out a livelihood
in the mountain belt, just below the snow-line ;
and it is a low-growing type, which must flower
early, or else it would soon be overshadowed by
taller rivals. For growth is rapid in the Alps,
once the snow has melted. Soldanella has thus
to blossom, and to secure the aid of its insect
fertilisers, at the precise moment when they emerge
from their cocoons in the first warm days of the
short alpine summer. If it waited later it would
be overtopped and obscured in a very few days
by the dense and rapid growth of waving grasses,
and aspiring globe - flowers, and long - stalked,
bulbous plants that crowd all around it. So
the soldanella seizes its one chance in life at the
earliest possible moment, and makes haste to pierce
its way through the solid ice-sheet, while lazier
rivals passively await its melting. That alone has

secured its survival and success in the crowded
world of the alpine pastures. For you must
not forget that while to you and me the Alps
are an unpeopled solitude, to the alpine plants
they are a veritable London of competing life-
types.

The canny plant lays its plans deep, too, and
begins well beforehand. It has made preparations. All the previous summer it has been
spreading its round leaves to the mountain sun,
and laying by material for next year's flowering
season. Leaves, you know, are the mouths and
stomachs of plants ; and the soldanella has a type
of leaves admirably adapted to its peculiar purpose : expanded in the sunlight, they eat carbon
and hydrogen the live-long summer, and turn the
combined oxygen loose upon the air under the
influence of the sun. By the time winter comes,
they are thick and leathery, filled with fuel for
the spring, and, of course, evergreen. They have
also long stalks, which enable them during the
summer to stretch up to the light ; but in autumn
they descend and flatten themselves against the
soil, so as not to be crushed by the snows of
winter. The first of my illustrations (No. 1)
shows a group of these fat leaves, seen from
above, and flattened against the ground in expectation of the snow-sheet.

The material laid by in the thickened leaves
consists of starches, protoplasm, and other rich
foodstuffs. The snow falls, and the leaves, protected by their hard and leathery covering, re-

main unhurt by it.   The food and fuel they have gathered is stored partly in the foliage and partly in the swollen underground root-stock.   All winter through, the plant is thus hidden under a compact blanket of snow, which becomes gradually hard and ice-like by pressure.   But as soon as the spring sun begins to melt the surface at the lower

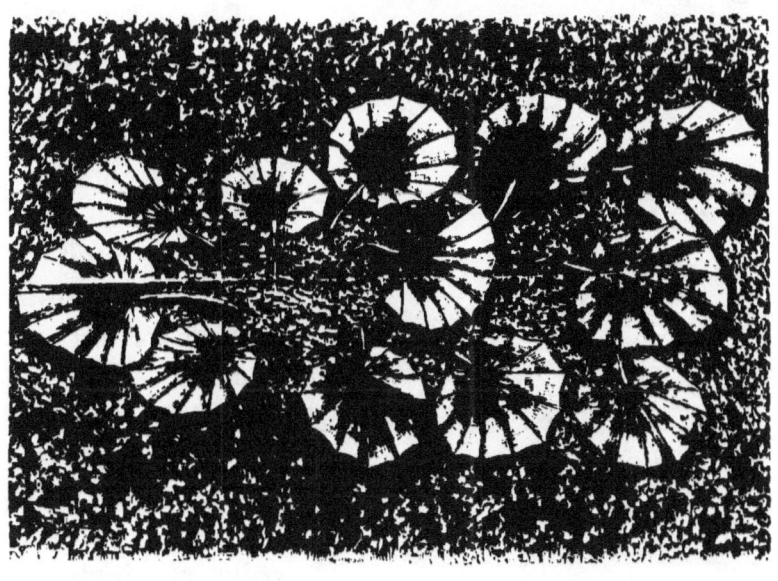

NO. I.—LEAVES OF SOLDANELLA IN AUTUMN, FAT WITH FUEL,
SEEN FROM ABOVE.

edge of the sheet, water trickles down through cracks in the ice, and sets the root-stock budding. It produces, in fact, the very same effect as the water which we pour upon malting barley in order to make it germinate.   And the same result follows, though here more definitely, for the sol-

C

danella has collected its material deliberately as fuel, and uses it up on purpose to melt its passage. It absorbs oxygen from the air below the snow, combines it with the fuels in its own substance, evolves heat from their combination, and begins to send up its nodding flower-buds through the icy sheet that spreads above it.

NO. 2.—BUD BEGINNING TO MELT ITS WAY UP THROUGH ICE IN A DOME-SHAPED HOLLOW.

The warmth the plant obtains by this curious process of slow internal combustion it first employs to melt a little round hole in the ice for its arched flower-buds (No. 2). At the beginning, the hollow which is formed above each pair of buds is hemispherical or dome-shaped; the stem pushes its way up through a dome of air enclosed in the ice; and the water it liberates trickles down to the root, thus helping to supply moisture for further growth with its consequent heating. But by-and-by the stem lengthens, and the bud is raised to a considerable height by its continuous growth.

Still, so slight is the total quantity of heat the poor little plant can evolve with all its efforts, that by the time the stem is an inch or two long, the lower part of the tunnel has curiously frozen over again, by the process which Tyndall called "regelation," and whose importance in glacier action he so fully demonstrated. In this stage, then, the melted space is no longer a dome; it assumes the form of a little balloon or round bubble of air, surrounding the flower-bud. At the same time, the ice beneath, having frozen again, almost touches the stem, so that the bud seems to occupy a small, clear

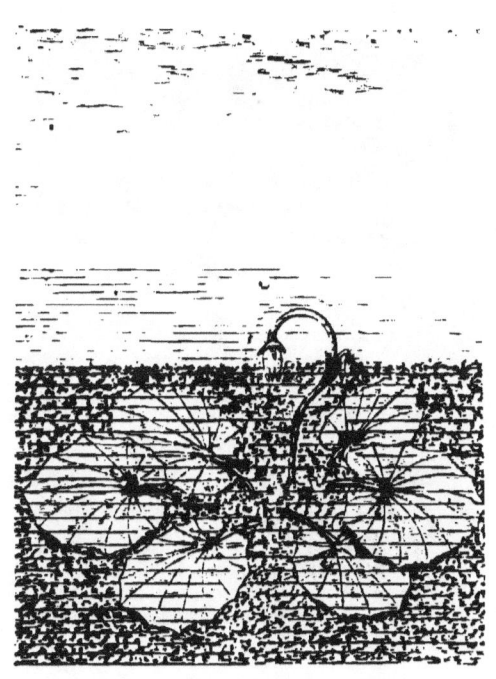

NO. 3.—BUD, SOMEWHAT LATER, EN-CLOSED IN A GLOBE OF AIR WITHIN THE ICE-SHEET.

area of its own in the midst of the sheet, with ice above, below, and all around it (No. 3). You would say that growth under such circumstances, in almost icy-cold air, was impossible—but if you examine the ice-sheet at the edge of the *névé*, you will find it studded by hundreds of such bubbles, each enclosing an uninjured soldanella bud in its centre. The reason is that the heat from the flower keeps the enclosed air just above freezing-point ; and so long as it is not actually frozen soldanella is indifferent to the cold of its surroundings.

NO. 4.—FLOWER REACHING THE SUR-FACE OF THE ICE AND OPENING IN A CUP-SHAPED DEPRESSION.

Gradually, in this way, the little buds manage to bore their way to the surface and to the sunshine on the outside of the ice-sheet. At last the stalk melts its path out, and a flower appears on the top, in the centre of a small cup-shaped or saucer-shaped de-pression (No. 4). The exquisite blue bells are thus seen bloom-ing in profusion, apparently out of the ice itself, or as if stuck into it. Unless you looked close, and noticed that their stems came from the ground beneath, you might even imagine they were rooted in the crystal mass of the *névé*. The edge of the snow-field in early spring is often pierced and riddled by hundreds of such soldanella borings; others above are in process of formation; and if you cut a piece open you will see inside how each is produced, with its narrow tunnel below, its balloon in the centre, or later, its saucer-shaped depression on the surface. Moreover, if you look at the foliage on the bare ground beneath, you will find that, when the flowers open, the leaves are no longer thick and swollen. All the fuel they contained has by this time been burned up for warmth; all the formative material has been duly employed in making the buds or blossoms, with the stems that raised them; and no-

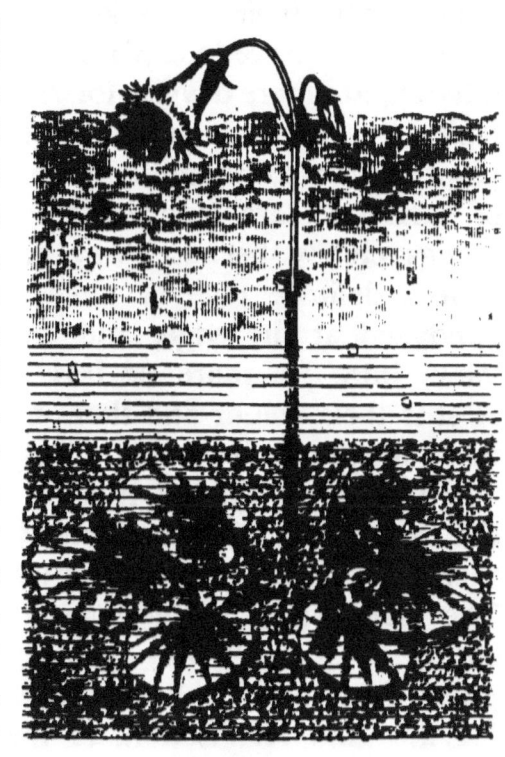

NO. 5.—FLOWER VISITED BY A BEE, WHICH FERTILISES IT.

thing now remains but drained and flaccid skeletons from which every particle of living matter has been withdrawn and utilised. Later on new leaves are produced in turn from the root-stock, after the ice has melted; and these new leaves, raising themselves on their long stalks, and catching the sunlight, begin afresh to accumulate material for next year's growth and next year's burning.

But why do the flowers want so much to reach the open air at all ?   Why should they not blossom contentedly under the enclosing ice-sheet? A glance at No. 6 will serve to explain the reason.   Flowers, after all, are mere devices for the fertilisation of the fruit ; it is the seeds and the next generation that the plant itself is mainly thinking about.   The blossoms of soldanella are noticeable to us lordly human beings chiefly because they are so pretty ; they have a delicate blue or violet corolla, exquisitely vandyked at the edge, and divided (on a closer view) into five more or less conspicuous lobes ; so it is their colour and their daintiness that make us so much admire them.   But to soldanella itself —which, after all, has to earn its livelihood with difficulty on a stern and rocky soil—this beauty that charms us is a mere matter of advertisement. The plant wants its blossoms to attract the early spring bees and honey-sucking flies, which carry pollen from head to head, and so fertilise its seeds for it.   And fertilisation, to the practical-minded plant, is the whole root of the question.   It cares no more for the beauty of its flowers in themselves than the British manufacturer of cocoa or soap cares for the gorgeous colours and striking designs he lavishes on his advertisements.   " Use Jones's Detergent " is the key-note of the poster.   The object of an advertisement is to catch the eye and secure the money of customers ; the object of the flowers, for all their beauty, is just equally to catch the eye and secure the visits of the fertilising insects.

No. 5 shows how all this is managed.   At the

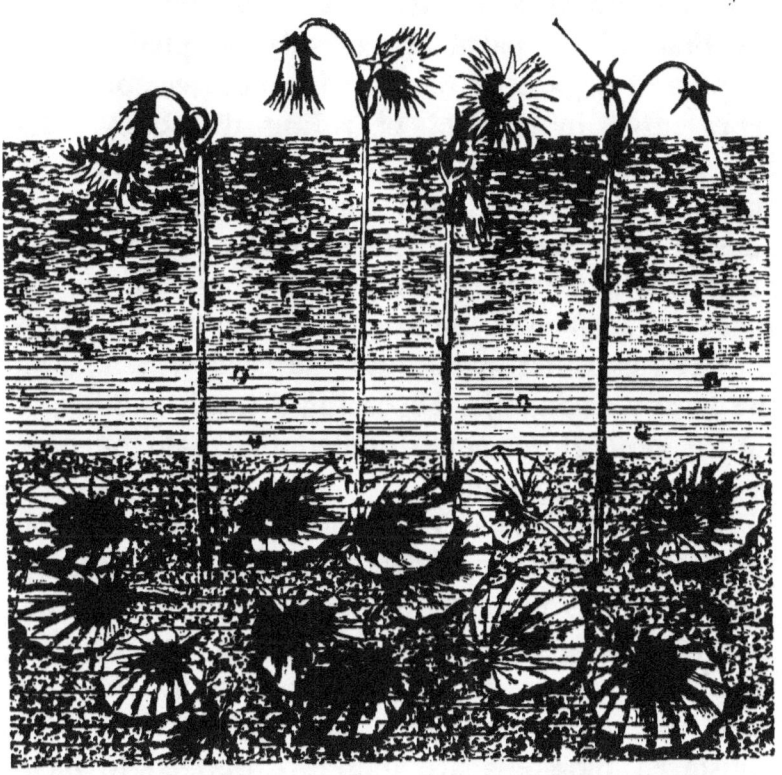

NO. 6.—GROUP OF FLOWERS IN DIFFERENT STAGES PROTRUDING THROUGH THE ICE-SHEET.

very same time that the soldanella raises its timid flowers, the bees and flies a little lower down the mountain sides are just escaping from their cocoons as full-fledged winged insects.   It is for their sakes alone that the pensive blossoms tint themselves in blue or violet, for you will find throughout nature that blue is the true bee colour ; and flowers that depend most for fertilisation on bees or their allies are almost always decked out in blue or purple. If you examine a soldanella closely, too, you will see that all its parts are exactly adapted to the shape and organs of its most frequent visitor, here represented in the act of rifling its honey. Its bell-shaped blossoms just fit the insect in size ; its stamens shed pollen just where his hairy body is adapted to receive it ; its sensitive stigma is so arranged that he rubs the golden grains off on the receptive surface of the next flower he visits. Then the little capsules swell, and the seeds ripen ; and the happy soldanella, becoming a fertile mother of future generations, has fulfilled the main purpose of its stormy existence.

Sometimes, however, the ice-sheet above is too thick to pierce ; and then the bud, after making manful efforts to melt its way out to the open air, is forced to give up the attempt in despair, and unfold its petals within its icy cavern. In that case, of course, no insect can visit it ; and such cloistered blossoms are therefore obliged to have recourse to the inferior expedient of self-fertilisa-tion. I say inferior, because all higher plants strive as far as possible to produce seedlings which shall be the offspring of a distinct father and mother. The last illustration (No. 7) shows two flowers which have lengthened their stalk in vain to the furthest point for which they possess material, but have failed to melt a way out of the solid ice-sheet.

They are therefore driven to curl round the tips of their stamens and fertilise themselves—a process which almost always produces inferior seeds and very weak seedlings. It is in order to prevent such disastrous results on a large scale, and to avoid the evils of constant "breeding in and in," that soldanella has invented its curious device for pushing its way boldly through its native ice-sheet to the sky and the insects. It goes there, not to look beautiful for you and me, but to secure the aid of its established pollen-carriers.

NO. 7.—PAIR OF FLOWERS WHICH HAVE FAILED TO REACH THE SURFACE, OPENING IN A SPHERE-SHAPED HOLLOW.

You must not suppose, however, that in doing all this the soldanella is displaying any extraordinary amount of unusual originality. Its speciality consists merely in the somewhat abnormal volume of heat which it generates. A great many plants, indeed, proceed much as the soldanella does in the matter of laying by materials for future growth in the leaves, and using these up in the act of flowering. Take, for example, the famous and often somewhat exaggerated case of the so-called "aloe," or American agave. It is commonly said that "the flowering of an aloe" takes place but once in a hundred years. This is a poetical fiction. As a matter of fact, the agave flowers on an average after fifteen or twenty years, and then dies down utterly. Every visitor to Italy or the Riviera knows this huge plant well—a gigantic house-leek in form, with its big spiny leaves and its points sharp as a needle, which defend it as by a bristling row of bayonets. Now, the agave lays by its material for future growth in the thickened base or lower portion of its leaves ; it thus forms a huge rosette, very much swollen and enlarged at the bottom. For years it goes on with exemplary patience, collecting supplies for its one act of flowering ; then at last, feeling its time has come, it suddenly sends up a huge stalk, or trunk, like a vast candelabrum, fifteen, twenty, or even thirty feet high, and supporting at its top a great bunch of big yellow blossoms. This enormous stem, with its colossal cluster of branching blossoms, takes only a few weeks to grow ; and as it rises and flowers, or still more as the immense capsules ripen their seeds, the bases of the leaves, once swollen and thick, become by degrees flaccid and empty. The stem and blossoms have drained them dry. At last, as the seeds fall, the whole plant dies away, having used itself up for ever in its one great act of flowering, just as the egg-laying rose-aphis uses itself up in its orgy of motherhood.

Now, this is much the same as the way in which soldanella behaves, except that soldanella continues to flower, spring after spring, for many years together. It does not exhaust itself in a single blossoming. Otherwise, the two plants, though so different in size, behave in much the same general fashion. For agave must necessarily evolve a great deal of heat during its rapid flowering period ; but this heat is useless to it, as heat, just as the heat we evolve in running a race is, as such, of no advantage to us. The main difference here is that soldanella has need of the heat and employs it deliberately for its own purposes. In the struggle for existence, every point of advantage any creature possesses must tell in its favour, and the soldanella has thus been enabled to hold its own bravely in the intermediate belt at the margin of the ice-field. But its limits are narrow. In the open ground it is soon lived down by more hardy kinds, which rise higher into the air ; its range is almost entirely bounded by a narrow belt just where the ice is melting. Above that point it cannot grow ; below it taller enemies soon oust and dispossess it. It utilises its short time between these two impossibilities.

Strange as it sounds, too, the ice itself acts as a sort of protective blanket or coverlet to the trustful soldanella. Only a plant that could pierce the ice could ever have hit upon such a paradoxical mode of warming itself by its own internal com-

bustion.   If a herb that flowers in the open were
to make experiments in warming itself in the same
manner, its attempt would necessarily fail, because
as fast as it heated the air the wind would blow
the heated portion away, and the plant would
therefore derive no benefit from its expenditure of
fuel.   But we all know how Esquimaux can live
in a snow hut, keeping it warm inside by their
own breath and the heat of their bodies.   It is
just the same in principle with the soldanella's ice-
cave.   The little dome or cavern gets warmed
within by the respiration of the flower-bud ; and
the heat thus produced is retained within the walls
of the cavity.   It is almost as though a mouse or
other small animal were to try to bore a path for
itself through an ice-barrier, not by gnawing the
ice, but by breathing upon it slowly till it melted.

See, then, how absolutely the soldanella behaves
like a man who is making a conservatory.   It lays
by fuel for the stove in its leaves to keep its flower-
buds warm and to force them in spring, at a time
when they could not blossom without the artificial
heat thus supplied them.   It keeps in this heat
within a transparent covering, the doors of which
are never opened.   As for light, that reaches it
through the crystal summit.   But it employs the
heat also to bore its way out ; and, as its ultimate
object is to get its young seeds fertilised, it finally
pushes its flowers out into the open air, where they
may receive the attentions of the fertilising insects—just as the gardener does, without knowing
why, when he wishes seed set.   The pendent bell-

shaped blossoms, again, even after they open, are admirably adapted for keeping in the heat ; and they are also exactly fitted to the shape and size of the bees and flies that act as their chartered carriers of pollen. A plant, in short, has to accommodate itself at every point to the needs of its situation ; it has to secure for itself a firm foothold in the soil, and a due share of food from the surrounding air (for its diet after all is chiefly gaseous) ; it has to take care that its pollen shall be duly dispersed, and its seedlets fertilised ; and finally, it has to see that its young are satisfactorily settled in the world, and deposited on likely spots where they can germinate to advantage. It must be a good parent as well as a prudent and cautious adventurer.

The struggle for life carried on under these circumstances has sharpened the wits of plants to a far higher degree than most people imagine. Plants have developed almost as many dodges and devices for securing food or avoiding enemies as animals themselves have ; and this single instance enables us to see with what forethought and cleverness they often provide against adverse chances. Soldanella, indeed, could not exist at all upon its ice-clad heights if it did not lay up food and fuel in summer against the needs of winter, like the bee and the ant ; if it did not burn up its own fat for warmth, like the dormouse ; if it did not tunnel the ice as the mole tunnels the earth ; if it did not retire beneath the snow-sheet on the approach of winter as the queen wasp retires into the shelter of the moss when frosts begin to kill her worker sisters, or as the squirrel retires into his hole in a tree at the approach of December. Ancestral instinct teaches the one just as much as it teaches the other ; and those who have closest watched the habits and manners of plants have the highest respect for their industry and intelligence.

Looked at from this point of view, we may consider indeed that every seed, bulb, or tuber is not merely a reservoir of material for future growth, but also a reservoir of fuel for supplying the heat necessary to the first stages of sprouting or germination. And without elaborating this question further, I may add that if you will examine closely many early spring buds and flowers, especially such as willow and hazel catkins, you will find not only that they are formed over winter and enclosed in warm overcoats to protect them from the cold, but also that they grow in spring before the air is warm enough to stimulate growth directly—or in other words, that they depend in part for heat on the consumption of their own internal fuels.

III

A BEAST OF PREY

THE lion, we all know, is the king of beasts ; a Tippoo Sahib of the desert, he treats his subjects with the simple and unaffected cruelty of an Oriental monarch. The tiger is also a somewhat ruthless animal ; he prefers to eat his dinner living. But for sheer ferocity and lust of blood, perhaps no creature on earth can equal that uncanny brute, the common garden spider. He is small, but he is savage. Lions and tigers are credited at least with the domestic virtues ; if we object to the king of beasts that (as Thersites said of Agamemnon) he devours his people, we may be told in extenuation that, like Charles I., he is a good husband and a model father. No such plea can be urged in mitigation of the misdeeds of that bloodthirsty wretch, the female spider. Not only does this Messalina among small deer poison, and then eat, her prey, but she also often kills and makes a meal upon her own lawful spouse, the father of her children. In selecting a garden spider of my acquaintance, therefore, as a theme for a short biography, I do not desire to hold her up to the young, the gay, the giddy, and the thoughtless as a pattern for imitation. She does not point a moral with the ant. On the contrary, she must rank with Semiramis and the famous queen who dwelt in the Tour de Nesle as a shining example of abandoned and shameless wickedness.

Spiders are not all alike. They are of many kinds, and of various families. So I shall begin by remarking that Rosalind, the particular lady whose portrait I have here presented to you in words, and whose life-history my colleague, Mr. Enock, has drawn for you from nature, belongs to the most familiar race of her kind, the true garden spider, which constructs the best-known and most perfect examples of regular geometrical webs. We called her Rosalind because she was a maiden of hunting proclivities, who lived under the greenwood in our own particular Forest of Arden. But her ways were not lovable. She killed flies in a fashion that would have brought up fresh tears in the eyes of Jacques; and she devoured her Orlando with all the callous ferocity of a South Sea Islander.

I will begin at the beginning with my eight-legged friend's biography. Rosalind was hatched in spring from a cosy cocoon or ball of eggs deposited by her affectionate, but otherwise cruel, mamma in the preceding October. She was one of a large family—say, seven or eight hundred. The cocoon was composed of yellowish silk, and attached, as the first illustration shows

you (No. 1), to the under side of a piece of trellis-work, against a cottage wall, partly overgrown with ivy. Within this snug abode the tiny eggs, each wrapped in its own internal coverlet, escaped the cold of winter, and hatched out in early spring with the first burst of warm sunshine. It was a bright May morning when they ventured abroad. The tiny spiders, just freed from their shell, with its outer great-coat, let themselves down by short webs to an

NO. I.—COCOON OF YOUNG SPIDERS HATCHING, AND SWARMING OF THE CLAN ON AN IVY-LEAF.

ivy-leaf below, where they clustered for a while, after the queer fashion of their species, in a sort of

D

close-knit crèche or communal nursery. Gathering
together in a compact ball or mass, like bees when
they swarm, the wee creatures began by spinning
in common a covering of thin silk, in whose midst
they lay rolled up in an apparently inextricable
tangle of legs and bodies. That is the universal
fashion of young spiders of this kind. But if you
touch them with a straw, a strange commotion
takes place all at once in the crowded home. The
mass unrolls itself. The six or eight hundred small
beasts within wake all together to a sense of their
responsibilities; the ball, which looks at first like a
cherry-stone, divides as if by magic into so many
eager and frightened animals ; and the spiderlings
disperse like the nations at Babel. Each goes his
or her own way helter-skelter, in search of a suitable place to commence operations as a general
flycatcher ; and in two minutes the space around
is fairly colonised by spiders, who set their snares
at once with exemplary industry. I am glad to be
able to give them credit for the one good quality
they do really possess ; though I am aware that in
their case industry is often only another name for
consummate greediness.

From the general gathering of the clan in which
our Rosalind thus took part she was rudely roused
by the touch of such a straw ; and, emerging in
haste into the open world, the great, cruel world,
amidst whose temptations henceforth she was to
earn her dishonest livelihood, she cast about her
for a favouring breeze to waft her first-spun threads
to some lucky position. It was a delicate operation.

Balancing herself with her eight legs on the edge of an ivy-leaf beside her native corner (as you see her graphically represented in No. 2), she span, to begin with, a few short ends of silk, which she exposed to a passing current of air by tilting her back up in her most persuasive manner. Where the silk came from, and how she managed to spin it, we will inquire hereafter ; for the moment, it must suffice to say that the wind was polite enough to fall in with her wishes, and to waft one of her threads to a secure position. There it gummed itself automatically by its own stickiness. Mr. Enock, who timed her, reports the interval she took in fixing this first thread as thirty-six seconds. The cable itself was drawn out from Rosalind's spinnerets by the force of the wind, as she stood with her head down and her body protruding; in little more than half a minute she was climbing up a line fifteen inches long, which had caught and glued itself on the edge of a jasmine leaf. For the silk is sticky and viscid, like the glue of a mistletoe, when first produced; it only hardens as it dries, so that it can be readily moored in its first state to whatever it touches. You may compare it in this respect to hot sealing-wax, or to the early pulled stage in toffee-making.

In No. 3, again, we see Rosalind's first snare, constructed neatly, with the usual architectural and geometrical skill of her race, between the twigs of the jasmine bush. In the centre she sits, as is her wont, head downward. The method of making this snare is so interesting and curious, however, that I shall describe it at some length, with needful explanations.

Rosalind began by letting the wind fix an original base thread, pretty much by accident. As soon as she was satisfied with the lie of this, she formed a

few others about it irregularly in a rough pentagon, as you see in the outer part of the web, merely to serve as a scaffolding for her future operations. But as soon as she had formed a careless angular figure all round the sphere of her projected snare, she let down a perpendicular thread from the top of her base, through the centre of her predestined home, and fastened it off at the bottom by gliding down it as she span it. Then, walking up

NO. 2 —-YOUNG SPIDERLINGS CASTING THEIR FIRST THREADS TO CATCH THE WIND.

this first ray-line again, she set to work once more a little to the right, spinning again as she walked, and fastened a second ray from the centre of the first to one of her outer cables. Next, time after time, she walked back to the centre, ran along the last ray made, trailing a thread as she went, and fastened each new line taut to one of the outer scaffoldings. So at last she had formed a regular set of rays like the spokes of a wheel, but as yet without any spiral connecting threads or mesh-like cross - pieces. The rays of this first framework were stout and thick, composed of several distinct strands, but very little viscid ; they were built up of many threads each, in a manner to be hereafter described ; and they hardened quickly on exposure to the air, for they were intended mainly to serve as beams, not as nets or insect-catchers.

Her ground-plan being thus complete, Rosalind next proceeded with great deliberation to add the meshes of the

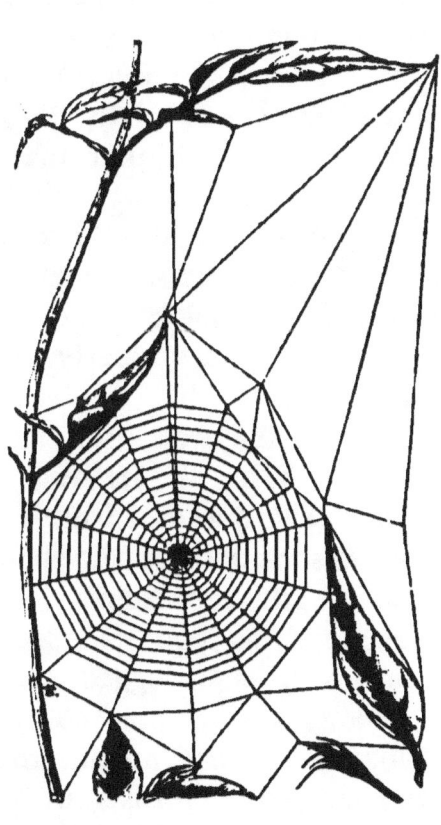

NO. 3.—A BABY SPIDER IN ITS FIRST SNARE.

web (which are the practical insect-catchers) by connecting the rays with the spiral network. In doing this, she followed a regular method. Beginning at the centre, she fastened a thinner cord to one of the spokes, and worked slowly outward, fixing the line to each ray as she went by the aid of her hind legs, which are almost hand-like. Then, reversing the process, she fastened another thread to one of the outer cables, and carried it back through the spokes in a similar spiral to the hub or centre. These two spiral threads are the ones which she specially designed for catching her prey ; they are thinner than the spokes, but are closely studded through all their length with tiny drops of sticky stuff like bird-lime, admirably adapted for snaring insects. You can see the drops, if you look close, even with the naked eye ; and they are very clearly visible by the aid of a pocket-lens.

How is the web itself manufactured and pro-duced ? What is its raw material ? Well, to answer that question I must give you here some brief description of the personal appearance of Rosalind and her sisters. The garden spider, you know (and as you can see her in No. 6), is a great, soft, eight-legged creature, about half an inch long, though her comparatively insignificant husband is very much smaller and less con-spicuous. She consists, in the main, of two parts, the foremost of which, though it rejoices in the scientific title of the cephalothorax (science is always so careful to give things nice easy names while it is about it !), may be more popularly described for most practical purposes as the head ; and to this large compound head are attached the eight long-jointed, hairy legs, with the muscles that move them. The other half of the spider consists of the abdomen or stomach, a soft, round bag, quaintly marked like a quail's head, and very squashy in appearance. With this last part of herself, the garden spider spins her snare or web out of the manufactured material of her own body. She spins it of her own digested contents. And as she has frequently to mend the web after various mishaps, which occur in the natural course of business—as when it is broken by the wind, brushed against by passers-by, or torn and mangled by a big fly or wasp—you can readily understand that she must eat in proportion; which is, no doubt, the true cause of her almost incredible voracity. In point of fact, a healthy female spider spends all her time in catching prey and eating it.

In No. 4 we have a greatly enlarged back view of the spinnerets from which the threads are produced, and a still more enlarged side-view below of the separate little ducts from which the component strands issue. According to circumstances, she makes her threads simple or compound. The sticky fluid of which they are formed is secreted by powerful glands in the abdomen ; it is then squeezed out through numerous minute tubes, of different calibres, and hardens in most cases when exposed to the air, though the spiral threads with the insect-catching drops on them maintain their viscid nature much longer, so as to gum the flies down, rather than entangle them in meshes, as with the common house-spider.

No. 5 shows us further details of some other interesting features in Rosalind's anatomy. The upper figure represents three distinct varieties of

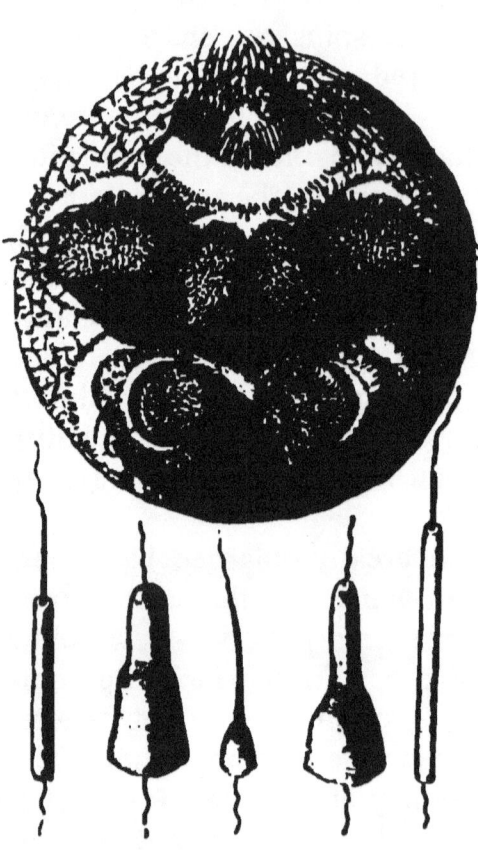

NO. 4.—BACK VIEW OF ROSALIND'S SPINNERETS.

the viscid threads, each with its own peculiar type of beads, adapted for catching larger or smaller insects. Every kind has its own beads spread for it. The flies get entangled in these, according to their size ; and then, tearing the web to free themselves, find the coils only double round their legs and bodies.

But the spider does not content herself with merely catching insects ; she poisons them as well. We had not watched Rosalind long in her chosen lair before we discovered that she did not live in her geometrical web ; that was merely her hunting-net ; her private residence consisted of a snug little cell or nest, under shelter of a rose-leaf, at a few inches' distance from the centre of the snare; and in this quiet home it was her habit to rest unseen, under cover of the shady leaf, until prey came within measur-able distance of her sphere of practical politics. But she kept up communi-cations with the seat of war. From the centre of the snare to the nest she had stretched a stout, thick line, along which she could run eas-ily on the slightest indication of a pro-spective victim looming up in the background. More-over, this cable or thread seemed to be connected by its dif-ferent strands with various parts of the snare; at any rate, it acted as a telegraphic

NO. 5. — VISCID THREADS, WITH STICKY BEADS; FOOT AND CLAWS OF SPIDER; SPIDER'S FACE, WITH JAWS AND POISON-FANGS.

communicator between the home, strictly so called, and the place of business. For Rosalind used always to recline at her ease with one hand-like claw placed steadily on the line of communication ; thus seated, she would watch with cat-like stealth for any chance of a victim. The moment a fly touched the snare, however lightly, it would set up a slight tremor of movement in the indicating thread ; and, quick as lightning, informed by touch of its whereabouts, out Rosalind would dart, ready to go straight to the spot and suck that luckless creature's life-blood.

Besides, the bigger the fly or bee, the harder it was likely to struggle ; and Rosalind noted well, before starting, the comparative extent to which the line was convulsed, and governed herself accordingly. If a big bumble-bee or wasp fell peradventure into her coils, he plunged exceedingly ; and Rosalind, prudently aware of the expected sting, approached the dangerous prey with marked reserve and caution. But when it was only a harmless small fly that struggled in the net, she rushed forth from her lair as bold as brass, seized the body with claws and jaws, and sucked the poor thing dry in less than a minute. Then she flung away its empty skin, or cut it contemptuously out of the web it had injured.

A glance at the second figure in No. 5 will show how admirably the spider's foot is adapted for all these various purposes. Adaptation could hardly go further. The spider has claws with which she can hold her web like a hand ; and she has also sharp nails which aid her not a little in manipulating her prey and her web. But she has more than all these : the claws themselves, you will note, are provided with toothed or comb-

like edges ; and these curious saw-teeth are useful
to the spider both in arranging her webs, in
weaving them tight or loose, and in feeling the
line of communication, when at rest, for indications of a captured insect.   If you remember that
the spider has no less than eight legs, each somewhat differently provided with special claws and
combs, you will understand how formidable a
beast she really is to creatures of her own size
or smaller.

But beneath the foot in No. 5 are represented
those still more terrible organs, the mouth and
poison-fang.   The face is shown, end on—a full-
face portrait ; and the little knobs above are the
eight sharp eyes with which the spider looks out
for its prey when captured.   Below lie the jaws,
with their two movable poison-fangs, one of which
is open, while the other is folded back into its
groove or receptacle like a kitten's claw.   This
poison-fang is supplied with venom from a gland
in the head.   When the spider catches an insect
and desires to eat him at once (as she generally
does if he is not very large) she poisons him outright, and proceeds to devour him.   So she often
does with a wasp or other dangerous insect.   But
if she wishes to preserve him for future use, she
quietly envelops him in a network of web, and
keeps him in durance vile, as I shall show you
later—a prisoner awaiting his turn to be killed
and eaten.   Taking her as a whole, therefore, the
mother spider is about as fiercely equipped a beast
as creation can produce : a monster armed like the

tiger and cobra combined; with the claws of a lion
and the poison-fangs of a serpent; both which she
supplements by a treacherous snare, itself a union
of the net and the bird-lime trap. No wonder,
with such an armoury, that she has prospered
exceedingly in the struggle for existence. And,
indeed, you will find garden spiders wherever you
go. They are one of the most successful types in
creation.

We watched our Rosalind closely through the
whole of a season. It was a curious drama of
blood and treachery. For the most part she lay
concealed like a secret assassin in her nest behind
the rose-leaf, seldom spreading her net in the sight
of the victim ; but sometimes, assuming the *rôle* of
highway robber, she would boldly rest in the very
centre of her snare, with her head downward,
waiting for the approach of casual small insects.
At such times, we noticed the larger and more intelligent flies usually gave her a wide berth ; she
seldom caught bluebottles or bees on these occasions of open display ; but tiny gnats and midges,
less careful or less wise, would get entangled in her
web, and at these she would rush out viciously,
sucking them dry then and there, and rejecting
their empty skeletons with lordly unconcern. Her
appetite was unbounded ; but she grew so quick,
she had so often to remake or repair her broken
snare, and she was laying by so constantly for her
maternal functions and her eight hundred eggs,
that this did not surprise us. The web, indeed,
was often torn by wasps or large flies out of all

recognition ; and at other times it was destroyed
by the housemaid or the gardener. On an average,
I should say, Rosalind had to rebuild the whole
concern about once in three days; and as she was
obliged to spin it all out of her own body, this came
very expensive. We noticed, however, that she
was economically minded, for she wasted no web ;
I think she ate up all loose ends or remnants : and
the central portion, where she occasionally reposed
on the look-out for prey, was free from the viscid
beads which elsewhere adorned the cross-pieces.
You see, this part of the structure was of comparatively small service as a snare, while the sticky
stuff would have interfered with her own freedom
of movement. She usually avoided the beaded
spiral, and only ran along the stouter spokes or
cables.

But the most wonderful scene of all was witnessed when Rosalind found in her net a large
wasp or a blow-fly. On such occasions, she was
generally resting in her nest under the rose-leaf,
with one foot held firmly on the cord of communication. If a light pull only came, she would rush
wildly forth, and seize in a frenzy the small fly
that caused it. She seemed as if drunk with lust
of carnage. But when the strength of the pull
showed her that a large bee or wasp was struggling
in the web, she would act in various ways according
to the needs of the moment. Wasps she approached, we noticed, with considerable fear ; she
knew their dangerous nature. But she was seldom
afraid, even so, of tackling them ; though at times,

if a very large and truculent specimen got entangled in the web, she seemed to despair of landing him. In such cases, she would cut him out bodily, by biting the threads, and let him drop at once, thankful, like Dogberry, to be rid of a knave. A moderate-sized wasp, however, she would rush out and attack in that frenzy of rage and hunger, a sort of mad, blind rage, which one often notices in fierce carnivorous animals. She would begin her onslaught near the victim's head, avoiding his sting, and envelop him in web, till his wings were pinioned ; then she would cautiously approach nearer and nearer to the tail, but give the actual sting a wide berth till the conclusion of opera-

NO. 6.—ROSALIND ON HER WAY TO SECURE A BLOW-FLY.

tions. The wasp, meanwhile, would keep protruding his poisoned lance in evident fury, striking wildly at the air ; while the spider continued to suck him dry quietly, from the head backward,

without the slightest consideration for his feelings
as a living animal. I may add (to anticipate an
obvious criticism) that I am aware the sting-bearing
wasp is a female; I have only treated her here
to a masculine pro-
noun because it helps
to discriminate her
better in each sen-
tence from my friend
Rosalind.

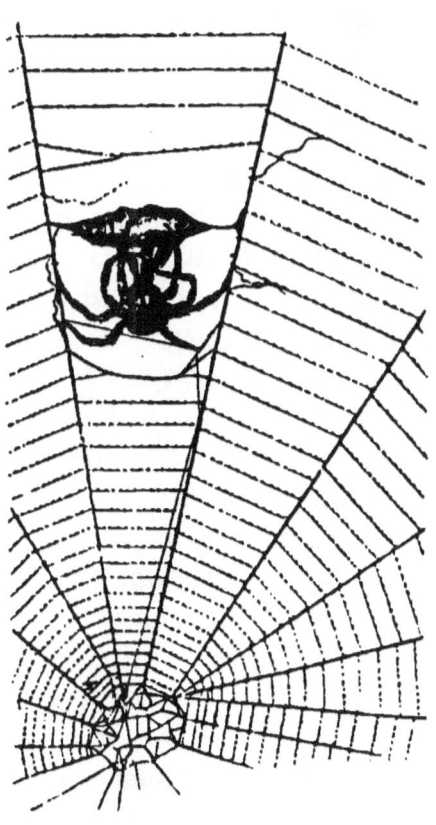

In No. 6, our in-
trepid Rosalind is re-
presented in the act
of attacking a blow-
fly which has buzzed
noisily into the web.
The moment her
delicate foot on the
line informs her that
a large insect has got
entangled in her toils,
she rushes angrily
out, and begins at
once to envelop him.
In this case, however,
her intention is not
to devour him on the
spot; she means to

NO. 7.—ROSALIND TRUNDLING THE
BLOW-FLY, AND ENVELOPING HIM
IN SILK FROM HER SPINNERETS.

store her larder with provisions for future use,
and is as careless as ever of the feelings of her
victim. No. 7 shows with what bands she proceeds
to swathe him. She catches him firmly as fast as

she can, so as to prevent his furious struggles from
unnecessarily destroying her precious web; then
she trundles and bundles him rapidly in a sort of
treadmill or merry-go-round, with her front pair
of legs; holds on to the web and steadies herself with her two middle pairs; and uses her hind
pair, with her comb-like claws, to distribute the
silk which she winds in coils about his wings
and body. You can see now how useful are her
eight legs to her. Each fulfils its own function.
In about a minute she has twirled him round and
round, and swaddled him firmly in a strong silken
covering. I regret to say she does not then proceed to eat him at once, but keeps him imprisoned
in torture for an indefinite period, tightly bound
in silken cords, till she desires to dine off him.
The unhappy fly is bound hand and foot—or,
rather, wing and leg—till it is absolutely incapable
of the least resistance; it is then kept in its close
prison with a cruelty more than mediæval, and
at last devoured alive piecemeal by its ruthless
captor. The morals of spiders are scarcely better
than those of Chinamen.

Rosalind's changes of costume were also most
theatrical and interesting. Like her namesake in
the play, she appeared every now and again
in a different suit of clothes, and rejected
her old ones. The manner of making the
new suit, however, and of shuffling off the old,
was extremely interesting. She moulted periodi-
cally; but at each moult the whole external
skeleton was sloughed off, like a snake's skin

or a lobster's coat, entire ; and a new one grew under it.

In No. 8 Mr. Enock has luckily caught our heroine just at the moment of such a moult. She is dropping out of her old skin, by means of her threads ; beneath it, the new one has grown, the animal being thus quite literally accommodated with a fresh suit "while you wait." The way the old skin hangs up is curious and typical. At first the new outer coat is soft and yielding, like the freshly moulted skeleton or armour of a crab or lobster ; but it soon hardens, and not infrequently advantage is taken of the moult to replace parts that have been accidentally lost or broken off, such as a leg or a feeler. The

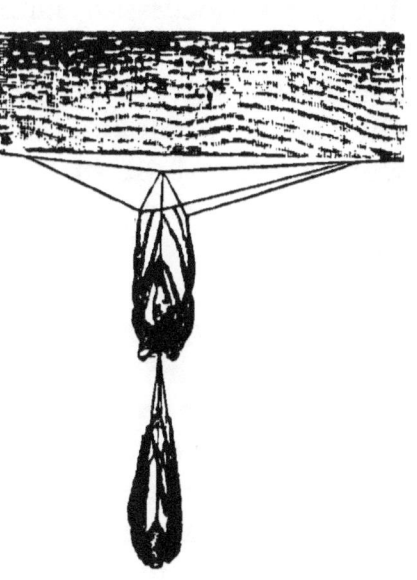

NO. 8.—A SPIDER CHANGING ITS SUIT OF CLOTHES.

economical spider, however, never wastes anything : she does not throw away the old suit ; as soon as her jaws have grown hard enough, it is eaten up by the owner, and thus used over again in the production of web or body material. If thrift be a virtue, no beast on earth possesses more than a spider.

E

I have left to the last the delicate question of
the domestic relations of spiders, which are certainly *not* of a sort to be commended for imitation.
The lady spider, indeed, too closely resembles the
late Mr. Deeming and the natives of Fiji in her
unsatisfactory notions of conjugal affection.   I
regret to say it is her reprehensible habit to devour
alive her unsuccessful suitors, and sometimes also
the father of her own children.   These are unamiable traits, but I must not conceal them.   You
will observe, no doubt, that throughout I have said
comparatively little of the masculine spider, and
much of his lady ; and I have done this of set
purpose ; for spiders are a group in which the
dominance of the females is marked and undeni-
able.   The matriarchate prevails ; the females are
the race, and the males exist only as lazy drones,
mere idle fathers of future generations.   This
being so, the mother spider, true to her thrifty
ideas, regards them in the light of necessary evils ;
and being always economical, she thinks it well to
utilise them for the purposes of the race by eating
them up the moment they have fulfilled their sole
and single marital function.

This peculiar habit makes the courtship of
spiders a grim tragi-comedy, well worth observing.   In No. 9 Mr. Enock has represented one
salient scene in the painful drama.   And this is
the interpretation thereof.   Two male spiders have
come to pay their court to the supercilious Rosalind.   She, good lady, sits unconcerned but watchful in the centre or hub of her snare, apparently

careless of the two eager postulants for her hand and heart, but in reality observing them with critical eyes, and ready to rush out and devour them if they fail to please her. The gentle-men, accordingly, have to be very artful. They go through strange antics. Now they approach her cau-tiously, very much on the alert, ready to pull the string and advertise her of their presence, but also prepared to turn and run, or to cut the line and drop, if she does not regard their advances with favour. Now again they retreat, alarmed at her aspect. Rosalind sulks in her web, and waits to see which of the two

NO. 9.—ROSALIND WATCHING HER TWO SUITORS, IN DOUBT WHETHER TO ACCEPT OR DEVOUR THEM.

she prefers, if either. Should the fit so seize her, she will accept one or other of her ardent suitors ; but should she happen to be hungry or else to be disappointed, or in an ill-humour, she may dart out upon them at once and make a meal off-hand of her devoted admirer.

Even the successful suitor himself is by no means safe ; for it is Rosalind's way, when she tires of a lover, not to nag and quarrel, but to devour him outright, and look out for another. This saves time and trouble, and is better in the end for the temper of the species.

When autumn comes, Rosalind lays her eggs in a cocoon, and fastens them on the under side of a stone or piece of wood, where they hatch out in spring, and so the whole story of her life begins over again. She herself, meanwhile, retires to winter quarters, where she passes the cold months under shelter in a state of more or less torpidity. It is not known exactly how long a spider lives ; but they continue for at least two or three years, and probably much longer. We had Rosalind under examination for two successive summers.

The family to which Rosalind belongs, that of the geometrical spiders, may be placed at the very head of the whole spider order. Its webs are the most perfect in architecture, are the best planned as snares, and have a strict monopoly of the sticky beads, which help to entangle the prey, and which are also, under the microscope, most beautiful objects, decked in prismatic colours, and looking like so many iridescent opals. In shape and markings these spiders are also superior to the common run of eight-legged beasts, though they are certainly less beautiful than some of the lovely green and variegated semi-transparent field-spiders. It would not be going too far to say that the geometrical web-makers are the most advanced and civilised members of the entire group. For there are degrees of evolution among these hunting carnivores. Some of the least advanced kinds merely stalk or hunt down their prey on the open. These lower savages among the spider tribe lurk under stones or in the crevices of bark, and rush out at their victims, or spring upon them unawares. One may compare them to such low hunting human races as the natives of New Guinea or the North American Indians. Others, again, construct tubes, with or without trap-doors, and catch their prey more or less cunningly near the entrance. Yet others, once more, weave irregular webs, among leaves and twigs, or in the corners of rooms, and trust rather to mere meshes than to sticky substances. But the geometrical web-weavers, the most advanced of their kind, have learned by the experience of ages how to construct a regular snare, on a fixed ground-plan, and to supplement it by a singular trick of beady bird-lime.

Even among the geometrical web-weavers themselves, again, there are marked varieties of progress and culture. For some kinds have only three claws to each foot, while others have more ; and there are certain species which possess in addition a sort of opposable thumb, so that they can catch things as with a hand, feeling them all round, and grasping their threads as a sailor grasps a cable. Such opposable thumbs are always accompanied by high intelligence, as one sees in man,

in the monkeys, in the opossum, and in the
parrot.

Indeed, all round, it may be safely said that the
spiders as a group stand at the head of the animals
with jointed bodies ; and that the geometrical tribe
in particular stand at the head of all the spiders.
Nor must we consider that their cruelty and ferocity
put them out of court in this connection ; for man
himself, taking him in the mass, is one of the most
ruthless of animals ; and the bees, which by universal consent rank among the highest insects, are
the group which most universally slaughter their
own brothers, the drones, as soon as the community
has no further use for them. The fact is that
Nature as a whole is intensely utilitarian ; each
kind fights for its own hand alone, and regards as
little the feelings of other kinds as the fisherman
regards the feelings of herrings, or as the fishmonger minds the objection of lobsters to be
boiled alive for our human convenience. A race
that skins living eels at Billingsgate, and decks its
hats with egrets in Hyde Park, has no just ground
of complaint, after all, against my poor, misguided,
husband-eating Rosalind.

IV

A WOODLAND TRAGEDY

NATURE is rich in tragedies ; but somehow, the tragedies which are long familiar to us cease to be tragic. We accept them as merely picturesque little episodes in our daily existence. Nobody is astonished, for example, when a cat plays with a mouse before killing it ; nor when she teaches her attentive kittens how to let it go in sport, maimed and half dead ; it does not shock us when the poor dazed little beast, thinking the danger over, makes a wild burst for freedom, that she shows them how to pat it with one cruel paw and still further disable it. Facts like these are too common and too long known to appeal to us strongly. We note them with a very languid interest. But when people first learn some unfamiliar example of Nature's cruelty, I almost always find they are profoundly struck by it. The novelty of the case gives it vividness and makes it sink in deep. And I know no instance which impresses the ordinary observer so much at sight as the first time when, wandering accidentally through some peaceful copse or wood, he finds himself face to face with that hateful hoard, a butcher-bird's larder.

For what the cat does with the mouse for a few short moments, that the butcher-bird does with it through long lingering days and nights of agony. He impales his mouse alive on the stout thorn of some may-bush, and keeps it there, maimed but struggling, or slowly dying, for a week at a time, until he has need for it as food for himself or his family.

A clever artist devised a cover for one of our popular scientific papers many years ago, which enforces well the universality of this ceaseless struggle of kind against kind, each wholly regardless of the other's feelings. In the centre foreground, a fly flits airily over the surface of a river, searching for its mate in the full joy of existence. Beneath, a small fish jumps up at the fly, and seems in the very act of seizing and swallowing it. Behind and below, however, a pike lies grimly in wait for the small fish with open mouth ; but he is anticipated by a kingfisher, which snatches it from his jaws before they can close over it. In the background above, a hawk poises itself on even wings, ready to swoop down in triumph at last on the successful kingfisher. There you have the epic of animal life in brief ; you have only to throw in an angler on the bank, fishing for the pike with a live-bait of minnow, and an enthusiastic ornithologist pointing his fowling-piece at the rare species of hawk, in order to complete the whole cycle of slaughter. And observe that each actor in this drama of death is as careless as to the life he sacrifices and the pain he causes as the angler is careless as to the feelings of the minnow he impales upon his barbed hook, or the sportsman is careless as to the feelings of the happy birds he brings down with his cartridges.

Nevertheless, when we come across one page .in this vast mute tragedy of sentient life among the calm surroundings of a quiet wood, it always surprises us afresh ; and that is why I have chosen as a good illustrative case of this phase in nature my wicked old friend the shrike, or butcher-bird.

Externally, I do not know that there is any-thing about his personal appearance which might lead you to suppose he was much wickeder or fiercer than the remainder of his family. In costume and colouring he is quiet and demure, not to say almost quakerish. To be sure, there is a lurking gleam in the corner of his eye, when you get a close view of him, which betokens a crafty and cruel disposition ; while something about the peculiar curl at the tip of his beak seems to suggest a lordly indifference to suffer-ing in others. But on the whole he is a hypo-crite in his outer dress ; you would hardly suspect him at first sight of the high crimes and mis-demeanours of which I admit him to be really guilty. Still, you do not know a thrush till you have seen him eat worms alive slowly, a mouthful at a time, pulling them out of their holes and chewing them gradually as he goes; and you do not know a butcher-bird till you have lighted upon him at home in his woodland haunts, with his living and writhing larder collected all round him.

In size, the butcher-bird (No. 1) is about as large as a lark; but he is a stouter and handsomer

NO. 1.—THE BUTCHER-BIRD.

bird, especially in his fresh spring plum-. age, when he goes a-courting, and wins his soberer bride by the beauty of his coat and the gallan- try of his bearing. His colouring is fine, but somewhat diffi- cult to describe, his recognised specific name of "the red- backed shrike" being perhaps too strong for his actual hues.

Chestnut, shading into reddish brown above, would be a more accurate mode of stating the facts; but he is pinky-white below, and has dashes of blue, of grey, of pure white, and of black scattered about in various parts of his plumage. A bright black bill and a dark hazel eye add beauty to his sharp and vigorous countenance. Alertness, indeed, is the keynote of his character.

As in most dominant races, his lady differs

much from him. She is duller and darker, and lacks the occasional white patches that adorn her lord. But she shares his general air of keen life and his rapidity of movement, being in every respect a helpmeet for him.

Mr. Enock has represented her in No. 2 in a characteristic attitude, perched on a small twig

NO. 2.—THE BUTCHER-BIRD'S WIFE.

of hawthorn, and ready to pounce down upon a luckless fly, whose movements she is watching with interested attention.

I say hawthorn on purpose, for the peculiarity of the butcher-bird is that in England or abroad it haunts for the most part thorn-bearing bushes. With us, it is but a summer migrant, occurring pretty frequently in the southern counties; but its winter home is on the Upper Nile and in East and South Africa, where it can find in abundance the thorny shrubs of the desert ranges, which

NO. 3.—PART OF THE BUTCHER-BIRD'S LARDER.

stand it in good stead as pegs or hooks on which to base its larder. In England, it usually selects a hawthorn for its scene of operations.

No. 3 shows far better than I can describe it

the nature of these food-stores, where the butcher-bird lays by meat for himself, his mate, and his unfledged young. The larder is always situated in the neighbourhood of the nest, and the male bird hunts for flies, bees, and other insects, while the female sits on the eggs hard by. He eats a few at once, to allay his hunger, spitting them first as a means of holding them ; but the greater number he preserves alive upon the cruel thorns for the use of his mate and his callow nestlings. "*Les pères de famille,*" said Talleyrand, "*sont capables de tout.*" And we may well exclaim, "Oh, parental affection, what crimes are perpetrated in thy name !"

The particular portion of the larder which Mr. Enock has selected for representation contains a bumble-bee, two large flies, and a nestling hedge-sparrow, stolen from its mother ; for the butcher-bird does not wholly confine himself to a diet of insects ; he is cannibal enough to catch and eat other birds, not to mention mice and such small mammals. So fierce and savage is he when on the hunt after provender, that he will even spear and impale larger birds than himself, such as blackbirds and thrushes. Not content with hanging them on the thorns alive, he will fasten down their legs and wings by an ingenious cross arrangement of twigs and branches, so as to prevent them from escaping ; for he does not so much desire to kill his prey, as to keep it alive till he is ready to eat it or to distribute it to his family. He knows that dead birds soon decay ; and he

doesn't like his game high : but he also knows
that wounded birds will live on and keep quite
fresh for days together; so he is careful to disable without actually killing the creatures he
captures.

Among the animals I have seen in butcher-
birds' larders I may mention mice, shrews, lizards,
robins, tomtits, and sparrows ; among the smaller
birds he especially affects willow-wrens and chiff-
chaffs : but keepers tell me that they have even
found them seizing and spitting young partridges
and pheasants. Whether this is true or not I cannot say ; but the game-preserving interest certainly
looks upon shrikes with no friendly eye, and you
may sometimes see one hung up on a nail among
the jays and hawks and stoats and weasels on
the "keeper's trees," where the guardians of the
wood display the corpses or skins of evil-doers
as a terror to their like, much as mediæval kings
displayed the heads of traitors above the gates of
the city.

Oddly enough, however, these "keeper's trees"
themselves are favourite haunts and hawking-pitches
of the butcher-bird, who is so little deterred by the
supposed lesson that he uses them as convenient
places for catching insects. For, in spite of his
occasional carnivorous tastes, your shrike is at
heart, and in essence, an insect-eater. He adds a
mouse or a tit as an exceptional luxury. Now, he
knows that the owls and stoats hung up on the
keeper's rustic museum attract numbers of carrion
flies, and he therefore perches calmly on the

boughs above the mouldering remains of his own
slaughtered brother to await the insects that come

to devour him.
Then he darts
upon them with
something of
the fly-catcher's
eagerness, eating
them up at once,
or flying off with
them alive to im-
pale in his store-
house.

In No. 4 we
see the female
butcher-bird, on
her return from
a successful chase
after prey of
greater import-
ance. She has
caught a harvest-
mouse, the tini-
est and prettiest
of our Eng-
lish mammals,
and though with-
out a license
to hang game,
has threaded it

NO. 4.—THE BUTCHER-BIRD'S WIFE
IMPALING A HARVEST-MOUSE.

through the neck on a branch of hawthorn, as
a preliminary to eating it. This enables her to

hold it conveniently as on a fork or skewer while
she pecks at it. Sometimes you will find the mice
fastened through the body, and gnawing the twig
with their teeth in their prolonged agony. But
the butcher-bird takes no notice of their writhings
and their groans : she treats them with the indifference of a fishmonger to lobsters. It is her
business to provide for her own young, and she
does it as ruthlessly as if she were a civilised
human being.

The shrike's ordinary method of capturing prey
closely resembles that of the fly-catcher, to which,
however, it is not really related. The resemblance
is merely one of those due to similarity of habit.
Every well-conducted butcher-bird has a settled
perch or pitch on which he sits to watch and wait,
and to which he returns after each short excursion.
Flies and bees he catches on the wing, darting
down upon them suddenly with a swoop like a
kingfisher's ; but he also often takes them sitting,
especially when they are settled on a leaf or branch,
or are eating carrion. One of his most favourite
hunting-boxes is a telegraph wire, and he prefers
one that crosses the corner of a wood ; there he
will sit with his head held sapiently on one side,
keeping a sharp look-out from his beady brown
eyes in every direction. If a bee lights on a head
of clover, if a cockchafer stirs, if a mouse moves
in the grass, if a fledgeling thrush makes a first
unguarded attempt to fly—woe betide the poor
innocent ; our butcher-bird is upon him, with a
fierce darting beak, and in ten seconds more, his

writhing body adds to the store in the shrike's larder.

A good place and time to watch a butcher-bird at work is in a quiet field by a copse just after the mowing. But you must hide carefully. The short grass is then full of beetles, crickets, and grasshoppers, as well as of mice, shrews, and lizards, who can conceal themselves less easily than they were wont to do in the long hay before the cutting. At such times, hawks and owls make a fine livelihood in the fields ; but their habit is to hunt their quarry on the open. They hover and drop upon it. That is not the butcher-bird's plan ; he is a more cautious and secret foe ; he sits casually on his branch or his telegraph wire, with his head on one side, till his prey stirs visibly ; then he pounces on him from above, making a short excursion each time, and returning to rest on his accustomed position. When he catches a bird, and eats it at once, he begins by spitting it on a thorn : then he attacks the skull first, breaking it through and eating the brain, which is his favourite tit-bit. He also makes raids on the nests of other birds, and carries off the nestlings.

If you open the crop of a butcher-bird, the contents will show you that, in England at least, its main articles of diet consist of bees and flies, but especially of beetles. It is full of their hard wing-cases. Now, ornithologists have long noticed that the distribution of butcher-birds in the land is very capricious ; in one district they will be fairly numerous (though, at best, they are rare birds),

F

and in another, close by, they will be very uncommon or quite unknown. It is probable that
this relative frequency or scarcity depends upon
the distribution of their proper food-insects. Indeed, just as we all know that an "army fights
upon its stomach," so we are beginning to know
now that commissariat lies at the bottom of most
problems of animal life. I used to wonder on the
Riviera why trap-door spiders, with their long
tubular nests, were abundant in certain deep red
clay-banks, but wholly wanting in others, just as
sunny, just as soft, just as easy to tunnel ; till one
day it struck me that the spiderless banks were exposed now and then to the cold wind, the *mistral*,
and hence were naturally almost flyless. As a
matter of course, the spiders went where the flies
were to be found ; and these open banks, though
sunny and warm, were from the spider's point of
view mere Klondykes or Saharas.

It is just the same with the butcher-birds.
Beetles and bees frequent for the most part
warm, crumbling soils ; they are infrequent on
damp clays and chilly, marshy places. Sandstone
and chalk attract them ; on London clay or the
damp flats of the Weald they are few and far between. Hence, where the beetles are, there will
the shrikes be gathered together. They abound
(comparatively) in warm sandstone hills, but are
almost unknown in chilly clay districts. Not that
they mind the cold as such ; it is the question of
food that really affects them. So, too, with the
swallows and other long-winged insect-hawkers.

The swift flies very high, and lives on summer insects, which come out in July and August only ; so he arrives here late, and goes away again sometimes as early as the date of grouse-shooting. The house-martin, on the other hand, subsists on low-flying midges which surround houses; he therefore comes first of all his group, and goes away latest. The night-jar flits over fern-clad or heather-clad moors, and feeds almost entirely on certain night-flying beetles and moths ; hence he arrives when they hatch out from the cocoon, and flaps southward again on his big, overlapping wings as soon as they have disappeared or been mostly eaten. It is all a question of commissariat. Our early English kings had manors of their own in many parts of the country, in all of which supplies were laid up throughout the year for the royal table ; in due time, the king arrived with all his court, stopped a month or six weeks, ate up all that was provided for him, and then rode on with his hungry horde to the next royal manor. It is just the same with the birds ; they come and go as supplies are assured them. The shrike stops in England while bees and beetles last ; when provender fails, he is off on his own strong wings to Rhodesia.

No. 5 introduces us to another strange scene in the eternal epic of prey and slaughter. It shows us how beetle proposes, but shrike disposes. Here, parental feeling wars against parental feeling. A busy group of burying-beetles have lighted upon a dead field-mouse — itself hawked at, perhaps, and wounded by "a mousing owl," but not quite killed at the time, and now abandoned on the open. The burying-beetles, all agog, proceed to cover it with a layer of earth—not, indeed, out of such instinctive piety as that which induced

NO. 5.—BURYING-BEETLES AND FIELD-MOUSE.—BEETLE
PROPOSES, BUT SHRIKE DISPOSES.

the robin-redbreast and the wren in the story to cover the Babes in the Wood with mouldering leaves, but for a much more prosaic and practical, though none the less praiseworthy, motive. They want to lay their eggs in it, so that the maggots may have plenty to eat when they hatch out—for these burying-beetles are carrion-feeders, whose larvæ thrive on dead and decaying animals ; and they desire to bury the corpse in order to keep it intact for their own brood, without interference on the part of other and more powerful carrion-eaters. When successful, they cover the mouse entirely with mould, and thus leave their young supplied with a liberal diet.

But hidden among the greenery of a tree overhead, a cynical butcher-bird is calmly watching those insect sextons from the corner of his eye. As soon as enough of them have collected on the spot, he will swoop down upon their bodies unseen from above, and will carry them off to spike them on his own pet thorns for the benefit of his struggling young family. Thus does parental affection war unconsciously against parental affection. Each kind fights only for its own hand, and regards only the young of its own species. For as Tennyson says well in " Maud " :—

" Nature is one with rapine, a harm no preacher can heal ;
  The Mayfly is torn by the swallow, the sparrow speared
    by the shrike,
  And the whole little wood where I sit is a world of
    plunder and prey."

No. 6 shows us one member of the butcher-bird's young family, just hatched and fledged, in his streaky grey plumage, and beginning to go out upon the world for himself. He is trying to catch an insect on a thorn above him. It also suggests to us the appropriate moral that if you train up a butcher-bird in the way he should go, when he is old he will not depart from it. Lessons of cruelty are here imbibed—I cannot

NO. 6.—THE NAUGHTY BUTCHER-BIRD—"I WANT THAT FLY!"

truthfully say, "with his mother's milk," but at least from his father's and mother's example. While the mother-bird sits upon her nest (as you see her in No. 7), the little chicks are fed "by hand," so to speak, with captured insects. But as soon as they can fly a little, they come out and perch upon the twigs of the larder, that

NO. 7.—THE BUTCHER-BIRD'S WIFE SITTING ON HER NEST.

they may learn fly-catching by helping themselves to insects spitted on the thorns, where parental affection, however misguided, has placed them for that purpose. Thus they imbibe a taste for

living food from their earliest moments.   As Prior
long ago put it :—

> " Was ever Tartar fierce and cruel
> Upon the strength of water-gruel?
> But how restrain his rage and force
> When first he kills, then eats, his horse?"

What the butcher-bird requires in his place
of residence, then, is, above all things, easy access
to warm sandstone or limestone tracts, with plenty
of insects, lizards, mice, and small birds ; he also
needs an open common to hunt over, bushes and
trees on which to perch at watch, and clumps of
thorn-bearing shrubs to provide him with a larder.
There he builds his rude nest, one of the roughest
and most inartistic I know ; and there the mother
brings up her young in her own wicked fashion.
But though a rather shy bird, the shrike does not
wholly fear or shun civilisation; for the rich insect
population of our garden often attracts the wicked
pair ; and in July and August, when flies are rife
among the fruit-trees, they will bring their young
brood into the currant and gooseberry beds, and
teach the young idea how to shoot in the manner
proper to so carnivorous a species.

As a matter of evolution, the shrike's position
is a very interesting one.   For he is not exactly
a bird of prey—certainly he does not belong to
the hawk and eagle order.   His near relations are
all mere insect-eating birds ; but he has gone a
little beyond them in his carnivorous habits, by
adding mice, birds, and lizards to his diet.   His

great discovery, however, is his cruel device of using thorns for his larder ; this ingenious but hateful invention it is which has secured him a place in the struggle for existence. It is curious to note, too, how the habit has reacted on the bird's structure and appearance. He has acquired the quick eye and nervous alertness of a bird of prey, and has even grown like that higher group to some small extent in his beak and talons. He is a wonderfully plucky little fighter, too, both against his own kind and against other species.

Have you ever reflected how wonderfully varied and eventful is the life of such a migratory bird as this cruel butcher ? We human beings, who can only travel south in one of the crawling expresses misnamed *trains-de-luxe*, have little conception of the freedom and variety which every mere shrike can claim as its birthright. Let us follow one out briefly through its marvellous life-cycle.

It is hatched from a creamy-coloured and dappled egg in a nest in England. From four to six brothers or sisters occupy the home, and, indeed, to be strictly accurate, more than fill it. Everybody knows the old conundrum, "Why do birds in their little nests agree ?" with its quaintly sensible answer, "Because, if they didn't, they would fall out." Well, with the butcher-birds, that remark is literally accurate. The nest is a ragged and rickety structure, hardly big enough to hold the young as soon as they are fledged. It is built in the boughs of a thorn bush, and near it stands the well-stocked parental larder. The young butcher-bird, as soon as he can fly, is taught to eat insects from the family hoard, and later on to pick them up for himself on the wing in the open. He is usually hatched about the beginning of June; by the middle of July, his mamma and papa take him on the insect hunt into neighbouring gardens. In his early plumage, he takes after his mamma, but already shows some signs of the white tips and black markings which will distinguish him as a male bird in his adult existence.

Once abroad in the world, he grows apace; and this is necessary, because, about September, he will have to fly off with his affectionate parents on a long, forced journey to warmer winter quarters. Not, of course, that he minds the winter in itself; but the flies and beetles are gone; their sole representatives are now the eggs and chrysalids; mice and lizards have retired into winter quarters; no small birds are about in the unfledged condition where one gets a fair chance with them; and altogether there is nothing for it but to travel south and find more plentiful support in some warmer country.

So southward the family flits, when partridge shooting begins, first over Channel to France, and then on to the Mediterranean. But food is scarce even in Provence and Italy during the winter months; so our wise young shrike and his parents do not loiter about with the invalids and *flâneurs* at Cannes or Naples; they strike right across sea, *viâ* Sicily and Tunis, to the Nile Valley. Thence, anticipating Mr. Cecil Rhodes and disregardful of railways, they keep straight on, with glorious views of sea and mountain, past the Mahdi's land, till they arrive at the great lakes and British South Africa. At least, that is the course pursued by the greater number, though a few more original families (mostly Russian by birth) trend eastward towards the Persian Gulf, and winter, after the now fashionable manner, in India.

During his absence in the south, our shrike grows adult, and also puts on his fine spring colours (which are his courtship suit, intended to charm his prospective mate), just before his return in May to England, or rather to Europe ; for of course I do not mean to say that he necessarily comes back to his native country ; though there is reason to believe that most migratory birds do really return year after year to the same quarters. They have a summer residence, so to speak, in France or England, and a winter one by the banks of the Zambesi or the Indus. Most butcher-birds that visit Europe in the spring come fairly far north, nesting in Northern France, Southern England, Belgium, Holland, or Germany. Few nest on the Mediterranean, probably because the summer droughts in that arid tract are unfavourable to their food-insects ; those that remain in Southern Europe or Western Asia choose, as a rule, the cooler and moister mountain regions, such as the Balkans, the Greek hills, Armenia, and the Caucasus. The English residents fly back from their African home (where they now enjoy the blessings of British rule quite as fully as in Britain) well fattened on juicy southern insects, dressed in their courting dress, and ready for the serious business of settling in life, choosing a mate, and rearing a young family. Indeed, observers in Eastern Africa have noted them during the intermediate period, sitting on the thorny shrubs, such as the Egyptian acacia, which abound in that region, and already adorned in their brilliant breeding plumage in anticipation of their return to their northern quarters.

Some people say that the shrike even makes two nests a year (as the swallow certainly does), one in the north and one in Africa ; but this is unlikely, and Dr. Sharpe, of the British Museum, will have nothing to say to it.

It is at the mating season especially that you have a chance, if ever, of catching sight of the butcherbird himself, seated, all eagerness, on his look-out tower ; and enjoying life with the calm begotten of that fine old recipe—a bad heart and a good digestion. He sits and utters his amatory feelings now and again in an abrupt little "chuck, chuck," which is whipped out suddenly, with a jerk of the head sideways as an appropriate accompaniment. About the same time, too—say the beginning of June—you stand the best chance of coming upon one of the larders, all stocked with fresh meat ; for later in the year, when the young are well fledged, the shrike gives up its murderous practices a little, and takes its young on the prowl for themselves in orchards and gardens, in order to accustom them to the habit of catching prey. But I suspect my evil friend of often murdering for mere murder's sake, as generally happens with predatory animals ; they acquire a certain love for the chase as such, and even seem, as one may observe in cats, to delight in cruelty for the sensuous pleasure of inflicting pain on others. Your shrike has no inkling of a conscience. He does wrong boldly, with sublime indifference ; and believes himself to the end to be a model father, a tender husband, an ornament to society, and a useful citizen.

V

MARRIAGE AMONG THE CLOVERS

PLANTS marry and give in marriage just as truly as animals. They have their loves and their hatreds, their friendships and their enmities. The marriage customs of many among them are vastly interesting ; and yet, in spite of all the attention that has been given to the subject of recent years, comparatively few people are even now aware how quaintly they pair, how varied and curious are their matrimonial arrangements. Most of us, it is true, have heard by this time the bare facts of the case—that flowers are mainly fertilised by the visits of insects : many of us even know that in the majority of instances the little golden dust which we call pollen must be transferred from the hanging bags on one blossom to the sensitive surface of another, or else seed will never be set ; but not all of us are aware how intricate and how numerous are the minor devices by which each kind of plant effects this important object in its own fashion. I am going, therefore, in the present paper to describe briefly the marriage customs of two alone among our commonest clovers, which I shall adduce as specimens of the strange variety to be found within the limits of a single type.

To begin with, however, I propose to examine, as a mere introduction, a couple of flowers of a well-known and dainty hot-house begonia, which may help us to the comprehension of the more plebeian clover-heads. Proverbial philosophy has long since taught us that "the longest way round is the shortest way home"; and when I drag in the begonia, which has apparently so little connection with clover, and which is really about as unrelated to it by descent as two flowering plants can well be to one another, you may suspect that I do so for some sufficient reason. The fact is, begonias happen to be plants in which the differences of the sexes are exceptionally well marked, so that they may be apprehended with ease by the naked eye and by every observer, even the most casual. I advise those who have conservatories of their own to verify my statements in this matter on the specimens in their possession; for those who have not, Mr. Enock's excellent illustrations, which accompany this paper, will serve almost as well as the original objects.

Most cultivated begonias have the flowers on their branches arranged in groups or clusters of three, the central one of which is often a female, while the two outer blossoms are usually males. This is the ordinary plan, but it does not hold good of all the species, some of which, on the contrary, have only one male to each pair of females. Now, these male and female flowers are so very unlike in

form and structure, when you come to look into
them, that you would hardly believe they belonged
to the same plant if you did not find them growing
on one branch together.   They differ quite as

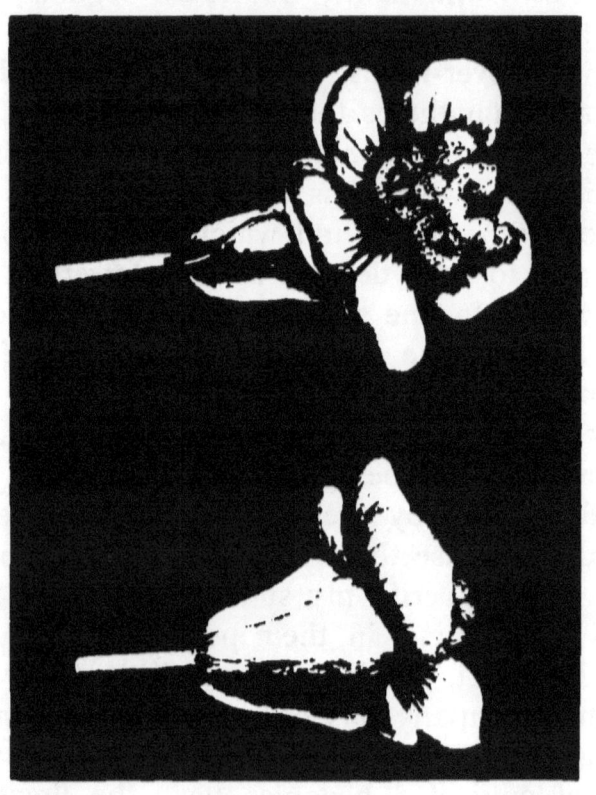

NO. 1.   FEMALE BEGONIA FLOWERS, FRONT AND BACK VIEW,
SHOWING THE SELF-BAG.

markedly as the peacock differs from the pea-hen,
much more markedly than man differs from woman.
A glance at No. 1, and then at No. 4, will make
this point obvious.   You would say, if shown them

separately, that these two blossoms must surely be flowers of quite distinct species ; yet they hang side by side on one and the same plant like brothers and sisters.

The first point of difference which you will note in the two is that the female begonia, as seen in No. 1, has five petals, while the male, in Nos. 4 and 5, has four only. (I call them petals both for brevity's sake and because I believe them to be so in reality, though fear of that terrible critic, Dr. Smelfungus, who goes about like a roaring lion seeking whom he may devour, compels me to add that in the learned Doctor's opinion they are parts of the calyx—a petty distinction with which, but for him, I would not have troubled you.) But what is far more important than the number of the petals is the fact that the female flower has wedged at its back a large triangular-winged ovary, or seed-capsule. It is the possession of this ovary, indeed, that marks it out at once as a female : for by a female plant or animal we mean, of course, the one which lays the eggs, produces the seeds, or becomes the mother of the young individuals. If you compare the back of the female flower in the lower portion of No. 1 with the back of the male flower in No 5, you will recognise at once the importance of this distinction. The female blossom has a seed-bag, while the male is barren. In No. 2 we have represented one such seed-bag cut open crosswise, so as to show both the projecting wings and the numerous little seeds in the three cells within.

G

But this is not all: the other parts of the two flowers differ almost equally. The centre of the female blossom is occupied, you will observe, by several twisted and wriggling arms, the upper surface of which is more or less sticky. This surface forms the receptive portion, or mouth of the flower, on which grains of pollen must be duly deposited before the embryo seeds in the capsule below can begin to swell and develop. On the other hand, the centre of the male flower, as seen in No. 4, is occupied by a set of very different organs, the stamens or pollen-bags, whose business it is to produce and shed the fertilising powder. Without pollen to start them, the seeds are useless. In the wild state, any winged insect which visits the plant is likely to alight first on the lip or platform of one or other of the outer male flowers. In his search for honey, which is secreted by the plant at the base of the petals on purpose to allure him, the flying visitor dusts himself over abundantly, though unconsciously, with grains of pollen from the very numerous little sacs which are placed there in a convenient situation with that

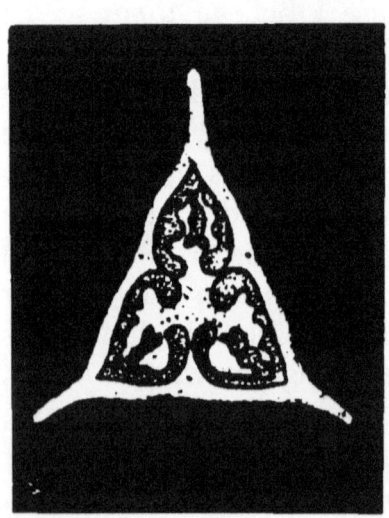

NO. 2.   THE SEED-BAG, CUT ACROSS.

precise object.   He then flies away to the female
flower, in which he alights, as a rule, on the central sticky portion (called by botanists the stigma):
and as he walks over it in search of the honey at
the base of each petal, he turns himself round and
round in five directions, and thus unwittingly
rubs off the pollen which clings to his legs and
hairs, transferring it to the sticky and receptive
surface.   After
visiting   and
fertilising   the
female flower in
the centre in this
manner, he then
usually   pro-
ceeds to visit the
second brother
beside it, from
which he carries
away pollen in
turn to the next
plant he visits.

NO. 3.—MALE BEGONIA FLOWERS IN THE BUD,
WITH NO SEED-BAG.

The object of this curious arrangement is that
each flower may be fertilised by pollen from
another blossom, and, as far as possible, in many
instances at least, by pollen from a distinct neighbouring plant.   But you will gather at once from
what I have said already that each plant must be
regarded in strictness not as an individual, but
rather as a community or commonwealth, of which
the leaves and flowers are the separate members
told off to perform different duties.   You may

compare it, indeed, to a hive of bees, the leaves representing the workers, while the five-petalled flowers are analogous to the queen-bees, and the four-petalled blossoms to the husbands or drones.

NO. 4   MALE BEGONIA FLOWER,
FRONT VIEW.

Nay, more: those of my readers who have begonia plants of their own may, observe for themselves another singular resemblance to the habits and manners of honey-bees. For after the drones have done their work in life by fertilising the queen-bee, the prudent workers sting them to death, as being useless mouths, of no further benefit to the community ; but the queen-bee necessarily survives to become the mother of young swarms, or future generations. If *she* were killed, it would be all up with the community.

Just so with the begonias ; as soon as the male flowers have performed their whole duty in life, by producing and disseminating the grains of pollen which the insects carry away and smear upon the sister blossoms, they break off at the

joint shown in the illustrations, and fall to the
ground; the plant refuses to feed them any longer,
because it has now no use for them: but the
fertilised female flowers remain fixed on their

stems to produce the
seeds, from which will
spring in time the
future generations.

What, however, do I
mean by fertilisation?
Well, each pollen-
grain, when closely
examined under a
microscope, looks like
a tiny egg, with a
very thin shell and
very sticky, active
contents. As soon as
the pollen-grains are
rubbed all over the
curly branches in the
centre of the female
flower, they empty
their contents down
long tubes, which
reach at last to the
seeds; and under this
vivifying influence, the

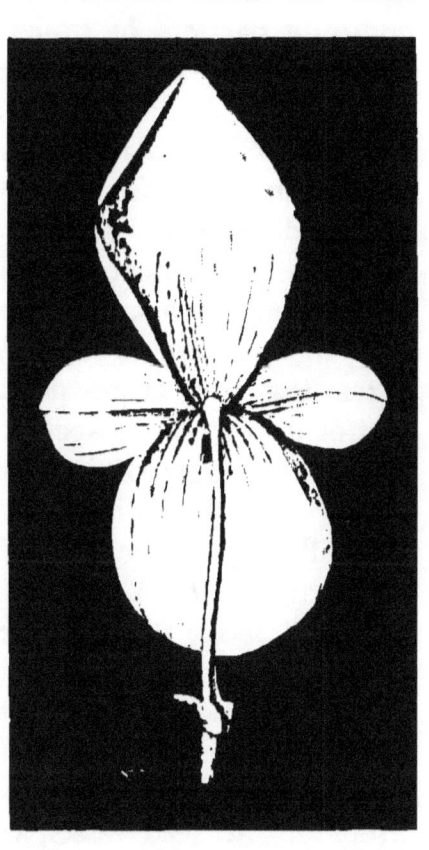

NO. 5.—MALE BEGONIA FLOWER,
BACK VIEW.

seeds begin to swell and become capable of producing young plants. The pollen, in short, has
quickening power. It is for the sake of this final
result alone that the flowers exist: they are pro-

vided with bright-coloured petals as advertisements to let the insects know where honey may
be expected; they secrete the sweet liquid itself
in order to induce their winged allies to become
common carriers of pollen for the benefit of the
begonia; and as soon as each flower has served
its purpose in this respect, it drops off or is
retained by the plant according as it is or is not
wanted in future for its seed-producing properties.

The difference between the brother and sister
flowers is even more visible in the bud than in the
fully opened blossom. No. 3 shows us this very
well in the case of an unopened male blossom.
Here the two large petals, afterwards used as
platforms for the insect to alight upon, enclose
the smaller pair of interior ones, as well as the
bunch of yellow stamens. But as these stamens
are full of nutriment, and therefore liable to be
prematurely attacked by useless gnawing insects,
the petal above them is thickened in this part, and
in one of the species most cultivated in our green-
houses, but not figured here, is provided with little
protective hairs, which baffle and keep at bay all
hungry aggressors. I may add that the projecting
wings on the seed-vessel, well seen in No. 1, and
also in the section in No. 2, serve a somewhat
similar purpose: they are intended to prevent
hostile insects from laying their eggs at the most
vulnerable points in the capsule, where the grubs
would destroy the seeds within. The thickenings
above and below, also to be observed in the lower
figure of No. 1, perform a like service. They

are devices of the mother to protect her young. You will thus perceive that the begonia has its friends and its enemies in the insect world, and that while it does its best to conciliate the one, it is no less anxious to repel the other. We shall find in the sequel that precisely the same thing is true of the clovers.

To the clovers then, which are our proper subject, I will next proceed. And I began with the begonia by way of introduction, only because that afforded us a case in which the husbands and wives of the community were so distinct from one another that nobody with a pair of eyes in his head could fail to distinguish them when they were once pointed out to him. In the clovers, on the other hand, we have a much more complicated arrangement, and one much less like the ordinary cases with which we are familiar in the animal world. Here, the flowers are collected in heads or clusters, and each flower is in itself at once both male and female. This method, indeed, is common amongst plants ; it occurs in by far the greater number of species: the reason why I started with the begonia is just because in that type the sexes are so well and clearly separated in distinct blossoms. In the clovers, however, each separate flower resembles a small pea-blossom in shape, having four petals, which botanists name respectively, from below upwards, the keel, the two wings, and the standard. These petals are best seen in the single upstanding flower (or " old maid ") represented in No. 9. They are enclosed beneath in a small

greenish calyx or cup, and they contain within them
ten stamens or pollen-bags, as well as a tiny capsule like a miniature pea-pod.   At the tip of this
capsule is a small hook—the sensitive surface on
which the pollen has to be deposited.   You would
say at first sight that under such circumstances,
male and females being mixed up in one, cross-
fertilisation must be impossible—that each flower
must surely be fertilised by its own pollen.   But
the clever clovers have invented an ingenious little
device of their own for overcoming this difficulty:
the pollen-bags and the sensitive surface of the
capsule do not arrive at maturity together.   In
this way each flower or plant gets fertilised itself
at one time by pollen from another plant, and at
another time dusts the bee that visits it with its
own pollen, which the bee transfers in due course
to the next plant it visits.[1]

No. 6 represents part of a plant of Dutch clover
—the common white clover of our meadows and
pastures.   It is called Dutch, not I believe because
it is particularly common in Holland more than in
other European countries, but because the prudent
Dutch were the first agriculturists to collect and
export the seed of this particular clover separated
from all other seeds of similar but less useful
species.   It happens to be a particularly good
fodder plant, and it grew wild originally throughout the whole of Europe and temperate Asia, from
the Mediterranean to the north of Norway.   But

---

[1] I hope technical botanists will forgive me some slight but unimportant simplifications in this not entirely accurate mode of presentation.

the seed has now been sown for pasture in almost
every country of the civilised world, so that wherever this volume circulates, its readers can find
and observe the plant for themselves, " to witness

NO. 6.—DUTCH CLOVER, BEFORE MARRIAGE.

if I lie," as Macaulay's Roman poet bluntly puts it.
Dutch clover is a rather smooth specimen of its
type, not nearly so hairy or silky as most other
clovers, for a reason which I will explain a little
later on : it has prostrate stems which creep along

the ground, as shown in the illustration, and root every now and again as they proceed, somewhat after the same fashion as strawberry-runners. Like all other clovers, it has trefoil leaves, each of the three leaflets in which is usually marked with a curved spot in the centre resembling a horse-shoe. But it is the flower-heads with which I am here particularly concerned. These are raised on long, erect, leafless stems, each of which bears on its summit a globular head of little white pea-flowers, often delicately tinged with pink or salmon. The flowers are thus lifted to a considerable height, because this clover grows, as a rule, among rather tall grasses, and so tries to push up its blossoms to a height where they may receive the polite attentions of passing insects.

The visitors for which Dutch clover specially lays itself out are for the most part bees. It disdains small pilferers. Each blossom has a long tube enclosing its honey, and only insects with a correspondingly long proboscis can reach its deep store of delicious nectar. It thus saves itself from being rifled uselessly by small insect riff-raff, such as flies and midges, which might visit the flower, as we botanists call it, "illegitimately"—that is to say, might rob the honey without conveying the pollen from the pollen-bags of one head to the sensitive surface or stigma of the next. The parts of the flower, in fact, are specially arranged with a definite relation to the head and the honey-sucking tube of hive bees and wild bees, which cannot visit it without dusting themselves over with pollen

on one blossom which they unconsciously rub off
on the receptive surface of the next.   In one word,
Dutch clover encourages bees for its own purposes,
because they are useful to it, while it places obstacles in the way of smaller and useless insects,
by burying its honey in a deep tube.

The head of Dutch clover shown in No. 6 is one
which has been caught just at the very first moment
of flowering.   The florets or blossoms which make
up the head begin opening from without and below,
inward and upward.   Thus in this head the outer
and lower florets have opened, while the inner and
upper ones are still in the bud.   When a bee visits
such a head of clover, he comes to it first from
another head of the same kind ; for bees do not
usually mix their liquors ; on one round of visits
they confine themselves, as a rule, to a single
species of flower only, and they probably store
the honey of each kind in separate cells, just as
we ourselves in our wine-cellars keep one bin for
champagne, another for claret, and a third for
Burgundy.   The bee thus begins with the outer
flower of the head, which he fertilises with pollen
from the last plant he visited ; he then goes on
to the second row, where he dusts himself over
with pollen for another flower-head ; and the buds
in the centre he leaves severely unnoticed.

As soon as he flies away, a very curious thing
begins to happen.   The flowers which he has
unconsciously fertilised close over their seed-vessel,
and grow gradually brown or withered.   At the
same time, as you see in No. 7, they turn down

out of the way of the bees by bending the separate little stalks on which they are raised in the head, and tucking themselves tight against the common

FIG. 7.　DUTCH CLOVER, THE FER-
TILISED FLOWERS TURNED
DOWN, THE UNFERTILISED
COURTING THE BEES.

flower stem. This they do partly in order not to confuse and worry their allies the bees, but partly also to avoid certain other dangers to which I will recur later. Plants often try in such ways to save bees or butterflies time and trouble, because the easier they make matters for the bee or butterfly, the more likely is he to visit and fertilise them. He is a useful customer whom they desire to conciliate. If a bee on his rounds finds that any particular species of plant gives him unnecessary trouble in getting at the honey, he is apt to neglect it and pass it by, in order to devote himself to other kinds which he sees are more business-like and obliging. The moment he comes to a head of Dutch clover, then he knows at once that he may safely ignore the dry brown flowers tucked away against the stem, because they are already fertilised and honeyless ;

he therefore directs all his attention to the mature and open flowers which are now producing honey and ready for fertilisation. These form practically, as you will see, at each moment the outer row of the flower-head, and are the ones which naturally first engage his notice as he alights on the cluster.

No. 8 shows us the same head in a little later stage of advancement. Here, almost all the flowers have now been fertilised, and they are therefore turning their brown and faded florets downward against the stem. Two among them, which the bee has only just left, are caught in the very act of bending

NO. 8.—DUTCH CLOVER, WITH ALMOST ALL THE FLOWERS FERTILISED, AND TWO JUST TURNING DOWN.

down, so as to get out of the way of any further visitor. The flowers in the centre, which are still erect, were not yet opened when the last bee paid a passing call on the community. They have unfolded their petals since, and are now standing up awaiting their turn to be visited by their winged aliy, relieved of their honey, and duly fertilised.

NO. 9. DUTCH CLOVER. A SOLITARY OLD MAID.

It sometimes takes four or five days for a single head to pass through all its stages.

In No. 9 we have a truly pathetic picture of a solitary old maid, perked up desolate and alone in the midst of her happier sisters. She was an unopened bud when some passing honey-gatherer visited and set the seeds of her more fortunate relations. The flower on her left, to be sure, has only just turned; it was the last to receive attention from its winged allies. If you search a field of Dutch clover, you will find every here and there such a solitary old maid. But you must bear in mind that none of this is true of the common purple clover, nor yet of the brilliant crimson kind (known to our farmers as "carna-

tion trifolium "), both of which are distinct species
with totally different marriage customs.   The ingenious habit of turning the fertilised flowers
downward out of the way of the insects is confined to a few species of
white, pink, and yellow
clovers.   It is a little
dodge on which they
happen to have hit, but
which has never oc-
curred to their larger
and more conspicuous
red and purple cousins.
So if you try to follow
out these hints in nature,
you must be careful to
hunt for white kinds
only.

No. 10 shows us the
last stage in the life-
history of a head of
Dutch clover.   All the
flowers have by this
time been fertilised; and
each flower alike is now
pressed down against
the stem in a crumpled,
brown, and withered-
looking mass.   The mere

NO. 10.—DUTCH CLOVER, ALL
THE FLOWERS FERTILISED,
AND MATURING THE SEED.

casual observer would say, "This clover is dead."
But it is nothing of the kind: it is only shamming.
The main object of the flowering and fertilisation,

after all, is the production of seed ; just as among birds the main object of pairing and nesting is the laying of eggs and the hatching of their little ones. And this introduces us to a second consideration of great importance. Plants take care of their young. The seeds of clover are small, but they are rich in foodstuffs laid by for the use of the little plant at its start in life. Now, the parent flower is well aware that many insects love to lay their eggs and hatch out their grubs in pods of this character ; if you have ever shelled peas, you must have seen such grubs very frequently in the pea-pods. The maternal instinct of the mother makes her lay her eggs where food is abundant ; the maternal instinct of the mother-plant makes it do its best to protect its young against such devouring enemies.

In No. 11 we see a flower of Dutch clover cut open lengthwise, so as to show the little pod within, very much magnified, and with one valve opened. Tiny as these pods are, they usually contain two, three, or four seeds. Every kind of clover, owing to the richness of these seeds, is much exposed to the attacks of insect enemies. To baffle these wary foes the clovers have invented an extraordinary variety of protective devices, two of which I mean to examine in this essay. Dutch clover meets the difficulty by tucking down the flowers after fertilisation out of the way of the bee, and then retaining the withered corolla or set of petals which completely enclose and hide the pod in the centre. Indeed, such a head as you see in

No. 10, all composed of brown and withered flowers, looks externally as if it were quite dead; but if you remove or cut open the sere and papery outer parts of the flower, you will find within them a vigorous little green pod, in which the miniature peas, after fertilisation, are maturing actively. In fact, the plant is only pretending to be dead; yet so effective is the pretence, and so well does the papery covering guard each pod against the egg-laying insects, that I cannot remember ever to have found a single grub in the seeds of clover. This may seem to you a small matter to guard against; but if you open the seed-capsules of the common little mouse-ear chick-

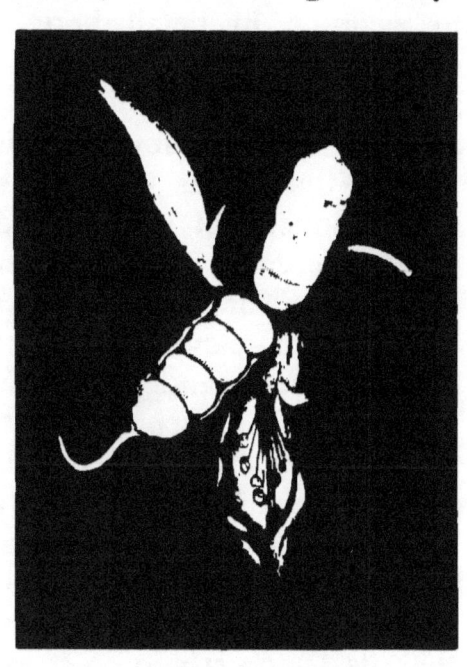

NO. 11.—DUTCH CLOVER, ONE DRY FLOWER CUT OPEN TO SHOW THE POD AND SEEDS RIPENING.

weed, which has no such protection, you will find in almost every capsule a small red grub busily employed in eating the seeds which the plant had laid by for the continuance of its species. It is thus a distinct advantage to the clovers in the struggle for life that they have invented devices

H

which enable them to guard their embryo young
from the assaults of insects.

Every species of clover—and there are many
—has some dodge of its own for thus protecting
its growing pods and seeds from the grubs which
would destroy them. I only propose, however,
to examine in detail here one more of these
dodges. We have another kind of clover, a
good deal like Dutch clover at a casual glance,
and commonly confounded with it by unobservant
people, though, as we shall soon see, the habits
and manners of the two kinds are in reality
very different. The strawberry clover, as it is
called, is a somewhat lower and smaller species
than Dutch clover, which it resembles in its
creeping stems and in its rich foliage. But the
flowers are not separately stalked in the head,
so that they cannot turn down after fertilisation
like those we have just been considering. Moreover, the stems and flower-heads are much hairier;
and this difference is due to the two facts that the
strawberry clover is smaller, and has a shorter tube
than its Dutch relation. It would thus be easy for
ants and other crawling insects to creep up the
stem and steal the honey, which is intended for
the use of fertilising visitors. To prevent this misfortune, and to keep its nectar for the regular
customers, the strawberry clover produces a number of hairs on the stem, which baffle the ants, to
whom such hairs are an impenetrable thicket. But
you may ask, " Why are not ants just as good as
bees for the clover ? " For this reason : flying

insects are mainly guided by sight and colour;
they flit straight from one flower to another of the
same species; and their heads are exactly adapted

NO. 12.—STRAWBERRY CLOVER, WITH FERTILISING BEE.

to the shape of the flowers, which in turn have
modelled their tubes and organs on purpose to fit
them. Ants and creeping insects, on the contrary,
are attracted merely by the sense of smell: they

notice scent of honey ; they climb up all stems indiscriminately in search of it ; they are bare-faced thieves with no organs adapted for carrying pollen ; and as they go about in the most reckless

No. 13. STRAWBERRY CLOVER BEGINNING TO SWELL.

fashion from one kind of plant to another, if they did ever by chance succeed in fertilising a casual flower, they would produce, not true species, but monstrous and meaningless hybrids. There-fore, many plants protect themselves by endless de-vices against the crawling ants, just as obviously as they endeavour to allure the winged bees, beetles, and butterflies. I may add that the head of strawberry clover is further protected against climbing insects by a num-ber of lobed bracts at its base, which effectually dis-perse these thieving ma-rauders.

While the strawberry clover is young and but recently opened, you might easily mistake it for a small and pinky specimen of Dutch clover. If you look closer, however, you will see that the petals are not so large, the tube not so deep, and the calyx much hairier. Nevertheless, as you may observe in No. 12, the hairs do not seriously get in the way of the bee during the stage when the flowers are just fit for ferti- lisation. As soon as the bee has left the plant, however, something happens which is quite different to the turning down of the florets in Dutch clover. The calyx or little cup which encloses each separate flower begins to swell and inflate itself like a balloon or bladder. In No. 13 you can see the beginnings of this curious process ; each calyx is slightly swelling round the tiny pod which it encloses. In Dutch clover, the pod is longer than the calyx, and the plant trusts for protection to the papery petals or corolla. But in strawberry clover, the calyx, after flowering, becomes very much inflated, thin, and netted; and in this state it completely encloses the growing pod. No. 14

No. 14.—STRAWBERRY CLOVER, AGAIN AN OLD MAID.

illustrates an intermediate stage in the process, with a solitary old maid still unfertilised, and the other flowers larger and more inflated. In No. 15 the inflation is complete : each little calyx has now swelled out into a small balloon, enclosing its pod.

The whole flower-head then becomes very compact, and assumes a pink tint, so that it somewhat resembles a strawberry, whence its ordinary name, though, as a matter of fact, it is much more like

a raspberry. You will observe that the beautiful network on the bladder-like head is closely covered with numerous hairs, which further help to protect the pods from the attacks of insects.

The truth is, Dutch clover is a denizen of rich and lush meadows, where it can take care of itself, and for which alone it is perfectly adapted. Strawberry clover, on the other hand, has chosen its home in close-cropped pastures, where its creeping habit and low stature help to save it from destruction. The dry and hairy heads are not relished by sheep, and you will often see them left uncropped where the neighbouring foliage has been closely nibbled. The swollen calyx with its hairs also keeps off egg-laying enemies. In No. 16 we have an illustration of one such fruiting flower, cut open lengthwise,

NO. 15.   STRAWBERRY CLOVER,
ALL THE FRUIT INFLATED.

so as to show the way the bladder-like calyx grows out around the pod as it ripens.

Now, what is oddest of all, every one of twenty or twenty-five species of clover has some dodge of its own for protecting its seeds after fertilisation. This shows how much these rich grains are sought after, and how carefully the plant is compelled to guard them. In some kinds, the calyx is a loose fluff of silky hair, enclosing the pod; in others, it is hard like a nut, or has stiff and pointed lobes which are sharp and prickly. One species closes its hardened lips over the growing seeds and pretends to be empty; a second develops a starry, thistle-like head, with tufts of thick hair, which conceal the swelling pod from observation. But the subterranean clover has

NO. 16.—STRAWBERRY CLOVER, A SINGLE INFLATED FLOWER CUT OPEN.

hit upon a still stranger and more ingenious device. It is a little creeping annual, much addicted to dry pastures or close-cropped hillsides, and particularly common on low knolls or barrows, nibbled over by numerous sheep and donkeys. Under these circumstances, it has a hard fight to protect its nutritious seeds and seedlings. It has taken, therefore, to producing small heads of loose white flowers, which look at first sight like poor specimens of Dutch clover. But if you gaze closer you will see that each tiny head consists of two or three properly developed flowers, with four or five undeveloped or abortive blossoms in the centre of the group. These undeveloped blossoms form a sort of living corkscrew. After fertilisation, the stems bend down towards the ground ; the corkscrew-like abortive flowers worm their way by pushing into the soil ; the pods are pressed down or buried in the loose mould ; and the plant thus sows its own seed for itself quite as effectually as a gardener could sow it. This is, perhaps, the furthest point which maternal solicitude has ever reached in the vegetable kingdom.

VI

THOSE HORRID EARWIGS

THIS is an age of vindications. Robespierre has been vindicated, and so has Marat; officious apologists have attempted to whitewash the unamiable character of Richard III.; Tiberius has been described as "a wise and great ruler"; and even poor Caligula has been lamely excused, on the ground of insanity, for such playful little freaks as making his favourite saddle-horse a Roman consul. Nobody's reputation is safe nowadays from the vindicator. It is the same in the animal world. New light is constantly being cast on the idiosyncrasies of the rattlesnake; we are assured from day to day that the cobra, though slightly venomous, is an excellent wife and a devoted mother; the scorpion only stings when you put him on the defensive or when he runs for his life; and the tarantula, we are told, has been most unjustifiably and cruelly blown upon. Has not the poet of "The Bad Boy's Book of Beasts" informed us that—

" The tiger, on the other hand, is kittenish and mild ;
   He makes a pretty plaything for any little child ;
   And mothers of large families (who claim to common sense)
   Will find a tiger well repay the trouble and expense."

In the midst of all these vindications, shall the harmless, unnecessary earwig go unvindicated from the aspersions that too often assail his character? A thousand times, no! Because he is small, he shall not be insulted with impunity. I see a helpless animal unduly exposed to vile detractions, and openly pursued with undeserved asperity. The sight arouses all the latent chivalry of my nature. I will gird on my sword to do battle for the right, and rush in, a scientific St. George, in defence of the innocent but persecuted earwig.

That my hero (or heroine) has a bad name in the world I am not careful to deny. Calumny has dogged it from its earliest days. Its very name enshrines a myth which is in itself a libel. It is called earwig, gossips will tell you, because it creeps into the ears of incautious sleepers in the open air, and so worms its way to the brain, where, if you will believe the purveyors of folk-lore natural history, it grows to a gigantic size, "as big as a goose's egg," and finally kills its unhappy victim. It is true, science knows nothing of this form of brain-disease ; it has tried the case before an impartial tribunal, and the earwig has left the court without a stain on its character. Some etymologists have even endeavoured to persuade us that the name earwig itself is but a corruption of ear-wing, a word which they suppose to be derived from the shape of its flying organs. There, however, our philologists are surely crediting the people with more knowledge than they possess ; very few gardeners or countrymen are aware that earwigs have

wings, while the general public never sees them
flying. Besides, the German name *Ohrwurm*, or
"ear-worm," and the French *Perce-oreille*, or
"pierce-ear," suffice to show that the myth is not
confined to our own country. All over the world
this harmless and on the whole beneficent creature
(for he is a good scavenger) is regarded with
superstitious fear and aversion ; all over the world
he is ruthlessly destroyed whenever found ; and
modern science alone is the first to attempt the
herculean task of rehabilitating him.

Before you begin to rehabilitate anybody, however, it is first desirable to know something about
himself, his family, and his antecedents. I will
therefore set out with a brief description of the
earwig and his relations. Almost everybody knows
well that earwigs are black little creeping insects,
which frequent dark spots, avoid the light, and
love to take refuge under stones or woodwork.
The earwig, in point of fact, is a nocturnal animal.
Like the bat and the owl, he hides during the
daytime, and only prowls forth at night in
search of food and adventures. Plain as he is to
outward view, his diet might suit the daintiest of
poets, for he lives for the most part on the petals
of flowers, on which account he is hated with a
deadly hatred by gardeners. But the diet of the
race is not wholly floral. Earwigs prefer petals
and other soft parts of plants ; but they will put
up with leaves or growing shoots, and even feed to
a small extent on dead or decaying animal matter.
That they are fond of fruit you must have observed

for yourself in the case of peaches and strawberries ;
though I fancy they never attack a perfect specimen for themselves. My own experience is that
they wait till a wasp has bored a hole in the rind
of an apricot or a nectarine, and then creep in
to enlarge it by their ad-
ditional efforts. If on any
such occasion, instead of
throwing the fruit away
in disgust, you will watch
the little robbers with a
pocket lens, you may (if
fortunate) have a chance
of observing the mode
of action of the mouth
organs. That is the difference between the point
of view of the naturalist
and the general public.
The outsider says : " What
a nuisance ! This peach
is full of earwigs !" The
naturalist says : " How
lucky ! Now I shall have
a chance of seeing how
he uses his mandibles !"

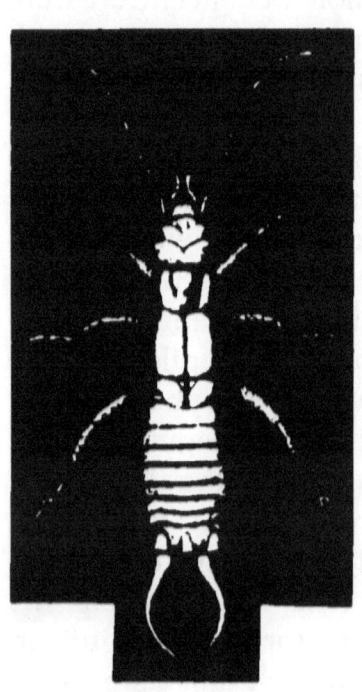

No. 1. PORTRAIT OF A GENTLE-
MAN. (OBSERVE HIS TAIL.)

And here let me call your attention in passing
to the portrait of a male earwig, the father of
a large family, in illustration No. 1. You will
observe at once for yourself that he has a long
body, divided as a whole into three well-demar-
cated portions. In front comes the head, with

its two beady-black compound eyes, its round
upper lip, its long waving antennæ, and its shorter
jaw-feelers. Next to the head come the three
rings or segments of the body proper (called,
technically, the *thorax*), each ring being here provided with a pair of legs, while the two hinder
rings bear also
wings or wing-
cases. Last of
all comes the
abdomen, or
tail, with its
numerous flex-
ible rings, of
which the male
has one more
than the female.
Notice also the
powerful pair of
pincers at the
extremity of the
tail, which are
the most con-
spicuous organs
in the full-grown
insect : they are

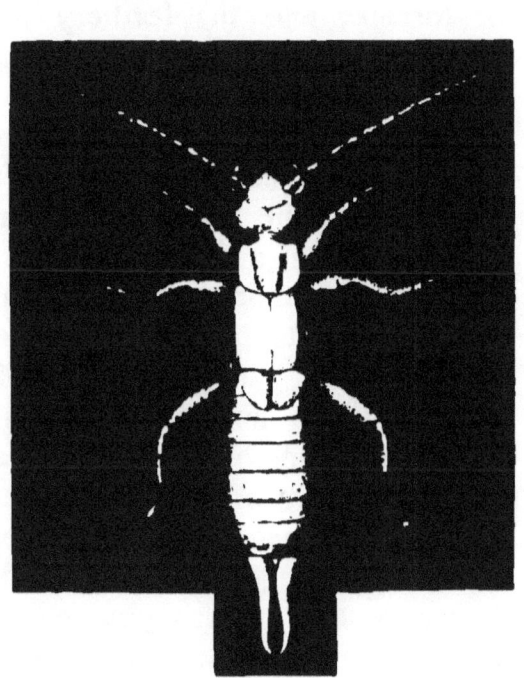

NO. 2.—PORTRAIT OF A LADY.

more curved in the father. of the family than in
his faithful spouse, and are likewise provided in
his case with curious teeth or indentations. The
use and meaning of all these parts will come
out in detail as we proceed with our inquiry ;
for the present, I will content myself with calling

your attention to the fact that "that horrid earwig" is a far handsomer animal when you come to examine him at close quarters than you were inclined to believe on a casual and disgusted summary inspection. Confess now that his beautifully jointed legs, his translucent thighs, his toothed pincers or forceps, and his feathery antennæ are very much finer than anything you expected from him when you first saw him.

In No. 2 Mr. Enock has given us the counterfeit presentment of the earwig's wife, for comparison with the portrait of her noble lord. You will observe at a glance that she differs from her mate in two main particulars only. She has one less segment to her tail; and her pincers, which are toothless, are almost straight and nearly parallel. The air of distinction which the husband thus gains over his wife is almost as marked as that which is given to man over woman by a couple of inches additional height, and by the noble appendage of a pair of black moustaches. Compare the two as you see them in the illustrations, and you will never again have a doubt as to the real nature of masculine superiority. If you are a man, indeed, I don't suppose you have ever had one. I have called the earwig black, but that is only true on a general survey. In reality, the head is rich chocolate brown, with the many-faceted compound black eyes standing out against it; the legs are amber-coloured, the jointed antennæ are pale amber, and the wing-cases are transparent or horn-like in colour.

Now, these two faithful portraits represent the earwig as we all best know him—the common or garden .earwig, engaged in crawling about during the hours of sunshine, and seeking some cranny where he may hide himself from the light that irks and distresses him. But there is another side to earwig life which in all probability you have never suspected. While day lasts the earwig shelters himself underground, or lies hid beneath stones or in the crevices of bark. But when night arrives, oh, then he sallies forth, on love and feasts inclined; he seeks his dusky mate, or battens on pink rose-petals. Then is the time to see him flying abroad on expanded wings; and then is the time when he really enjoys existence, till some late-flying swallow or prowling bat puts an end to his brief revels.

" But I never knew earwigs *flew!*" you exclaim. " I never thought they had wings. Those I have seen were always creeping and crawling."

That is quite true; and in this matter I will not deceive you. The common earwig does really fly; but he is an infrequent aeronaut. Indeed, I believe he seldom uses his wings except when he is courting or changing his residence. However, there is a smaller species of earwig, not minutely discriminated from the common sort by housewives and gardeners (who kill all the race impartially), but known to entomologists as *Labia minor*. This lesser member of the tribe may often be seen disporting himself on the wing on warm afternoons in summer; and even the larger earwig occasionally ventures out after dark in the same manner. The approved method of taking earwigs on the wing is by means of a tarred board, on which they may be caught in small numbers. When the broad transparent wings are expanded, they are really beautiful and striking objects.

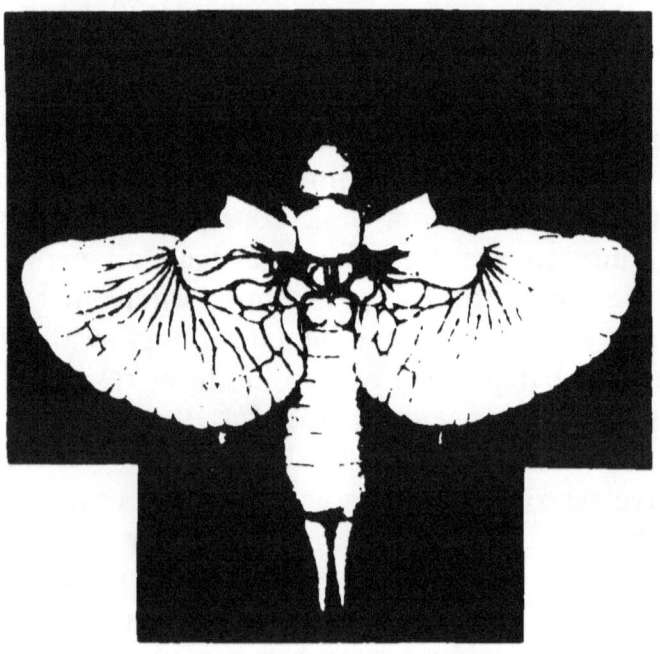

NO. 3   FEMALE EARWIG WITH HER WINGS EXPANDED.

What becomes of the wings, however, when the insect is at rest or crawling? Well, they are almost invisibly tucked up in a most curious and marvellous way under the horny outer pair, or wing-cases. In beetles, the horny front pair or wing-cases completely cover and hide the hind pair or flying wings. But earwigs are in many ways a less advanced and perfect group than the beetle tribe ; as we shall see hereafter, they are a rather primitive tribe, only half way up in the scale of development towards the highest insects. And among their imperfections one may mention

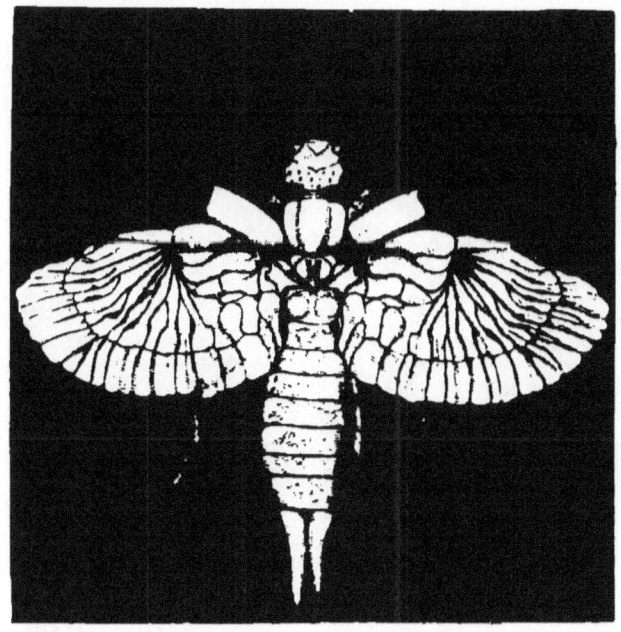

NO. 4.—BEGINNING TO CLOSE.

this—that the hind wings are only partially covered by the front pair or wing-cases.

When I say so, however, I do not mean to be unkind to the earwig, who, within his own limitations (as we say of minor poets), must be looked upon as one of the most marvellous and complicated of animals. And I propose to illus-

I

trate this fact for you in a single direction by
a brief consideration of the way in which he folds
and tucks away his pinions when he has done
with them.

No. 3 represents a female earwig in flight, with
the thin, transparent wings fully expanded.   You
will notice here
that the first
pair, or wing-
cases, which
are hard and
horny, are held
open in front
out of the way;
and that the
second pair, or
true wings, are
flat and papery
behind,   but
have a curious
horny rib or
"stiffener" in
their front por-
tion.   This stif-
fener acts ex-
actly like the

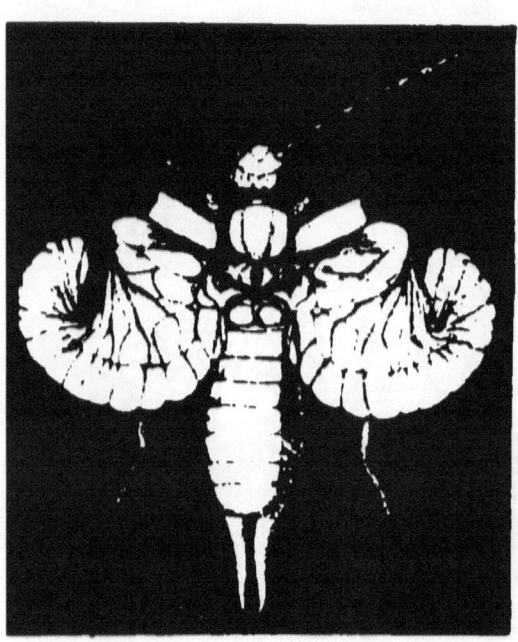

NO. 5.—DOUBLING UP THE FORE-WING
FANWISE.

whalebone or steel in a pair of corsets, or like
the ribs in an umbrella.   The beautiful folds
and creases in the true wings resemble those
in a fan or Japanese parasol; but they run two
ways, some lengthwise, and some transversely.
They are exquisitely true in their wrinkles, and

enable the insect to shut up the wing with perfect accuracy.

No. 4 and the subsequent illustrations show us the various stages in the very complicated closing process; and Mr. Enock has so drawn them for me as to let us follow in detail every step in this wonderful piece of insect jugglery. Cinquevalli himself does nothing more admirable. To see an earwig close her wings is a study in the perfection of Nature's mechanism. In No. 4 itself, which is the first of the series, the rib or stiffener is just slightly depressed, so as to make the tip of

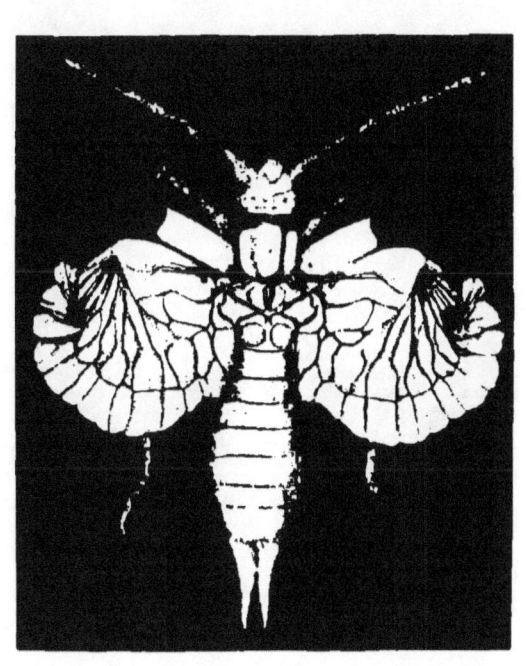

NO. 6.—A STAGE FURTHER.

the wing drop a little. In No. 5, the stiffener bends at the joint in the middle, and thus makes the edge of the wing curl inward like a fan, the pleats folding neatly with the utmost precision. With the stage illustrated in No. 6, the wing begins to flap; and in No. 7, the first part of it disappears round the corner, while the remainder turns up like a hinge at the intermediate cross-nerves. In No. 8, we find the wing constricted in the middle by the process of folding; while in No. 9, the back part has been nicely tucked away behind the front por-

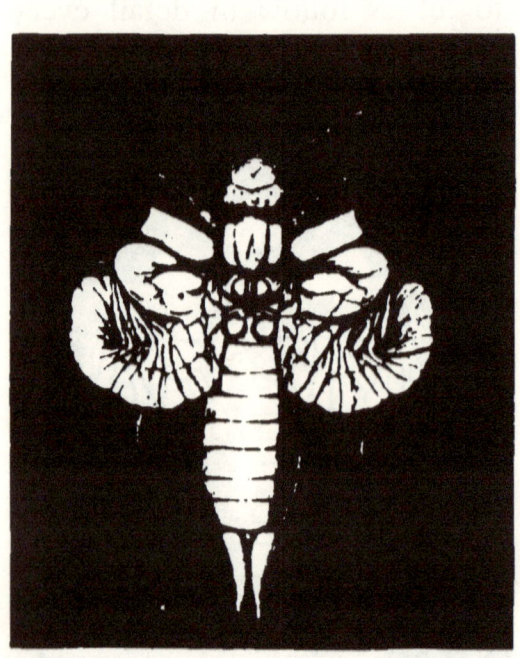

tion, so that the whole simulates for a moment a pair of separate wings. In Nos. 10 and 11, again, the folding still continues, till the muscles which move the wings have done as much as they can do in the way of tightening up, by their unaided efforts. And now comes in the use of the tail with its curious appendages;

NO. 7. THE BACK PART FOLDING HINGE-WISE.

and very odd it is. The pincers supplement the action of the wing-muscles.

As soon as the earwig has reached the point of closing represented in No. 11, she suddenly turns up her tail from behind, as you can see in No. 12, opens her forceps, and applies the sharp points of the pincers to the recalcitrant

wing-tip, which will not close of its own mere
motion. Then, as you can observe in No. 13,
she rapidly clips the pincers together, thus tucking
in the last bit of the wing much as a hand might
do it. After that, she straightens her body again,
as in No. 14,
and is ready
to replace the
folded wings be-
hind the hard
wing-covers. Of
course, all this
process, which
we have repre-
sented here in
detail in its vari-
ous stages, only
occupies in life
a few brief se-
conds; so per-
fect and so au-
tomatic is the
mechanism that
the earwig man-

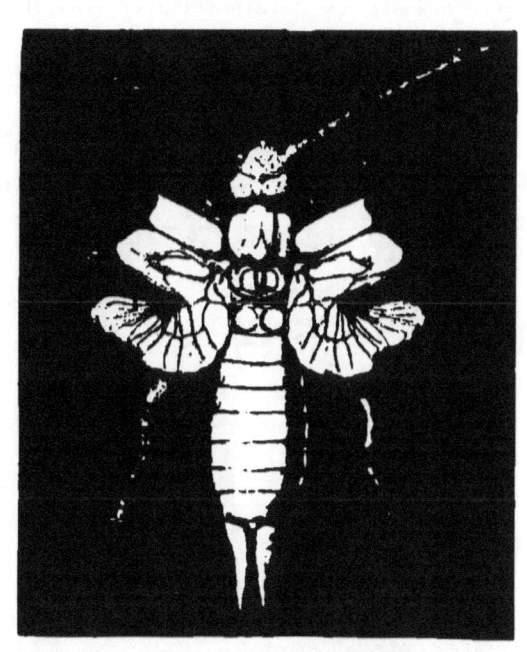

NO. 8.—A SECOND LATER.

ages it all as readily as a lady closes up her fan
and reopens it.

In No. 15, our earwig is shown in the act of
replacing the folded wings over the abdomen;
while the hard, horny wing-case is beginning to
cover them. In No. 16 she has folded them quite
back, but has lifted the wing-cases again, as if to
fly off once more; this illustration exhibits the

size of the wings when fully folded, and enables you to understand their true relation to the outer wing-cases. Reverting now to No. 2, the mechanism is seen once more completely closed up, and the earwig is prepared to crawl about on the ground in its usual sedate and humdrum manner.

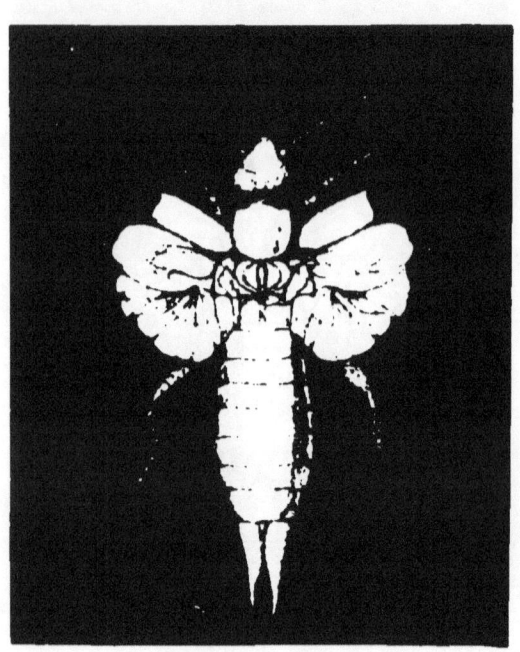

No. 9.—THE HIND PART FOLDS BENEATH THE FORE.

But if, after this, you ever despise those horrid earwigs, I shall think you have no taste for the wonderful in nature.

Perhaps, however, the most marvellous point in the history of the female earwig is the fact that she sits on her eggs and takes care of her young exactly as a hen does. She retires underground to lay her eggs, which she deposits in some safe and convenient cranny usually ready-made for her. She is not herself a good digger, like the mole-cricket, nor has she feet specially adapted for clearing away the soil; she therefore takes advantage of accidental cracks in the ground (being a cave-dweller, not an excavator), and is particularly fond of following the disused burrows of earth-worms. You must remember that the surface-soil is literally honeycombed with burrows of worms, which are not mere holes, but neat small tubes, cylindrical in outline, carefully engineered, and lined throughout with a layer of fine earth, as solid as concrete. The mouth of the burrow is also frequently papered with dead leaves, cemented to the wall by a sticky secretion from the worm's body. These underground tunnels often penetrate the earth to a

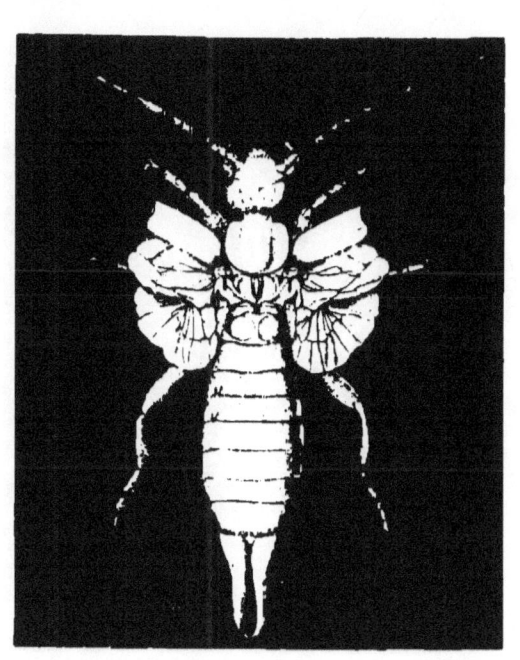

NO. 10.—THE PROCESS CONTINUED.

depth of many inches, and occasionally go down as much as six or seven feet. They thus form excellent approaches or adits, which the earwig can use in prospecting a suitable cranny for her own nursery. If you ask why the worm does not expel the intruder, or stick up a notice to say that trespassers will be prosecuted, I would point out in reply that hundreds of such tunnels are rendered tenantless each day by means of thrushes, starlings, and other worm eating birds, which prowl about lawns, gardens, and meadows, picking out the earthworms as fast as they show their noses above the level of the soil; while hundreds more are made desolate by moles and centipedes. There is thus never any lack of empty burrows which the earwig can appropriate, as the hermit-crab appropriates the empty shells of whelks and periwinkles.

NO. 14.—THE WINGS THEMSELVES CAN GO NO FURTHER; SO —

In No. 17 we see the mother earwig safely installed in a nice underground nest, and sitting like a hen on the eggs she has deposited within it. You can dig up such nests and eggs in any garden in January and February. Mr. Enock tells me he sometimes finds them at a depth of six inches. The average number of eggs in a brood runs from fifty to sixty. The good mother sits on them till they are all hatched out, and even then continues to watch them, as a hen does her chicks, till they have arrived at years, or rather weeks, of discretion.

No. 18 is a portrait of the earwig and her numerous family in their first condition. And this picture leads us up to one most interesting point in the earwig's development. You will notice here that the young insects closely resemble their mother in most respects — far more closely than a caterpillar resembles its butterfly ; they have the same sort of

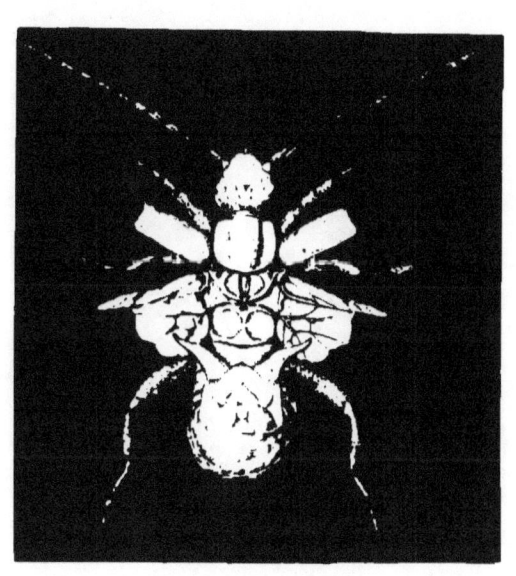

NO. 12.—THE TAIL COMES IN TO HELP THEM.

head, the same sort of body, the same sort of tail, and the same peculiar pincers ; but they are quite wingless. Now, this brings out in a very clear way their analogies to and their differences from most higher insects ; it enables us to form a distinct idea of the origin of that standing miracle, the metamorphosis of the maggot into the fly and of the caterpillar into the butterfly.

Some insects have wings, and some have none ;
but among insects with none, we may distinguish
two classes : those whose progenitors could fly,
but who have themselves degenerated so as to
become wingless ; and those who never had wings
at all, but represent the primitive non-flying ancestor. Several
of these early
wingless types
still persist to
the present day ;
and they very
closely resemble
the young of
the earwigs.
They have a
head with a
couple of wav-
ing antennæ ;
they have a body
of three seg-
ments, each of
which bears a
pair of legs, but no wings ; they have a long,
jointed abdomen ; and at its end they have two
appendages, which, though not specialised into
pincers, distinctly suggest the forceps of the earwig.
Indeed, if the baby earwigs always remained in
their first larval stage, we might easily mistake
them for some of these primitive wingless creatures. No. 19 is a rough sketch of such an early
type of non-flying insect, by name Campodea.

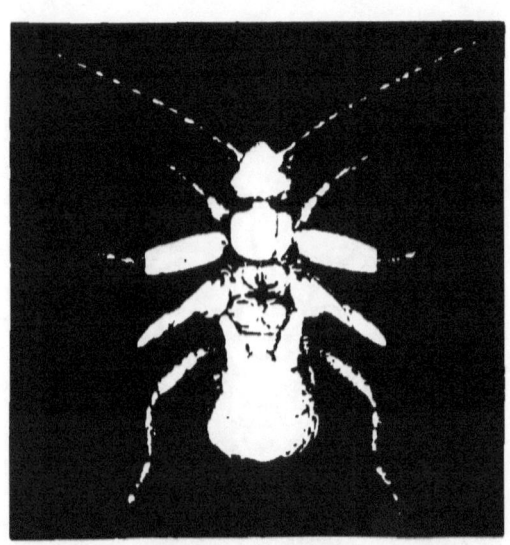

NO. 13. —THE USE OF THE PINCERS.

The young earwig, however, does not stop short at this point. When born or hatched from the egg, he closely resembles his parents in most respects, and as he grows and moults, he becomes at each change more and more like them, till at last he is justly considered "the very image of his father." At a certain stage in his development, indeed, we find that on two segments or rings of the body, two prominences or protuberances begin to make their appearance. These are the rudiments of the wings and wing-cases, which grow gradually under the skin, and be-

NO. 14.—THE TAIL STRAIGHTENED OUT AGAIN.

come fully developed after the last moulting. We may fairly take it for granted, therefore, that in this case the young earwig when first hatched out resembles the original wingless ancestor of the race ; but as time goes on, he begins to assume the various forms which the race has passed through in its advance to the modern winged condition. In other words, the metamorphosis of the individual sums up for us in brief the evolution of the kind.

Observe, however, that the young earwigs do not pass through any distinct and well-marked

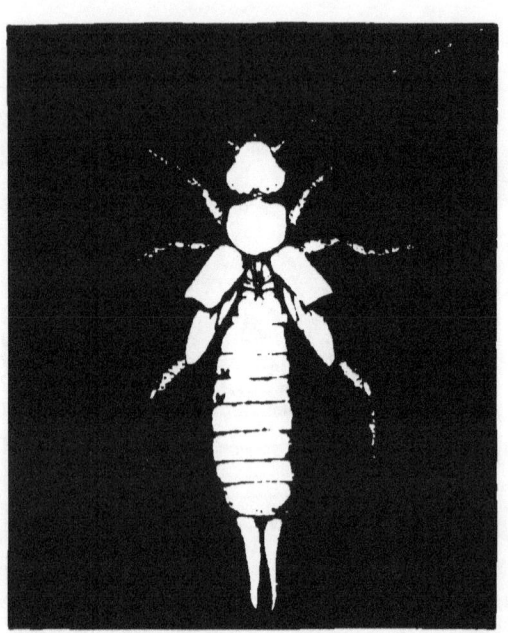

stages of larva, pupa, and imago grub, chrysalis, and butterfly like their more advanced relations. It is true, the names of larva and pupa are frequently given to the two earlier phases in the life of the earwig and its allies. But the terms are misapplied. All that happens to the earwig

NO. 15. REPLACING THE WINGS BENEATH THE WING CASES.

is a gradual series of successive moults; and during one of these moults the wings make their appearance. Moreover, the young earwig when just hatched out of the egg (as you can see in No. 18) resembles its mother in everything essential save in the possession of wings. There

is no real metamorphosis, or a very imperfect
one ; hardly more change, indeed, than takes
place in the growth of humanity ; for the acquisi-
tion of walking and the addition of a beard and
other adult adjuncts may fairly be compared to
the development of the wings in the growing earwig. It is quite
otherwise with
those insects
which under-
go a complete
metamorph-
osis, like bees
and butterflies.
The young
grub in the
comb does not
in the least re-
semble the full-
grown bee,
whether queen
or drone, or
worker ; the ca-
terpillar does
not in the
least resemble

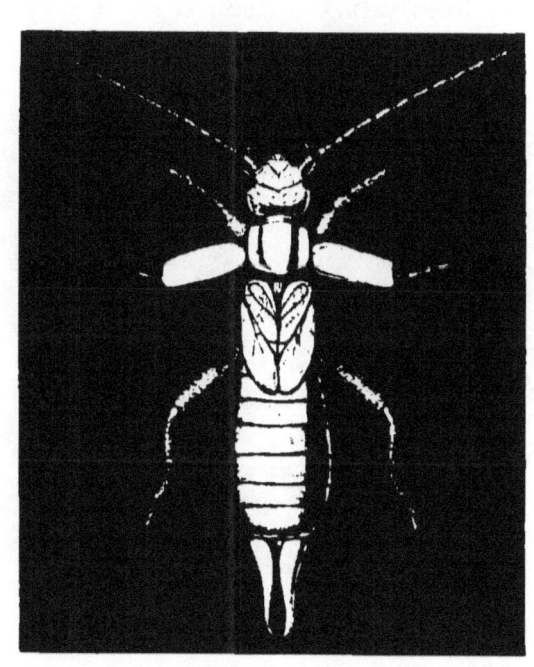

No. 16.—THE WINGS AT REST ; THE WING-
CASES RAISED AGAIN.

the beautiful full-grown moth or butterfly.

And here we get another curious piece of cross-
relationship ; for while the young earwig only
" throws back " to a primitive six-legged, wingless
insect, such as the one figured in No. 19, the
young bee or butterfly " throws back " to a far

earlier stage, and is hatched out in the form of a crawling worm—a type which must have belonged to a much more original ancestor. It passes the first stage of its life in this worm-like form, but it does not grow by slow degrees, like the earwig, into its final shape. On the contrary, it suddenly boxes itself up one day in a pupa-case,

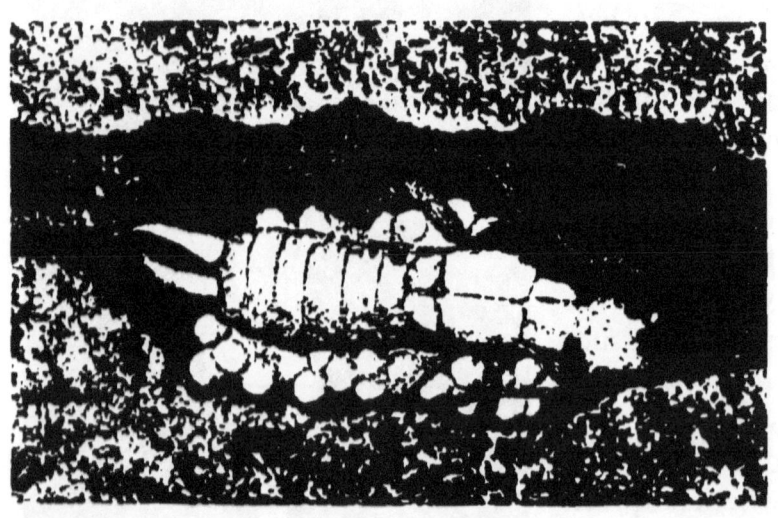

NO. 17.—THE MOTHER EARWIG SITTING ON HER EGGS.

or chrysalis, lies by dormant for a while, rearranges its parts entirely, and then rapidly develops into a wholly different creature—a bee or wasp, or moth or beetle. The earwig's change is growth; the butterfly's is a transformation scene.

How are we to explain these facts? I think in this way. Long, long ago, the common progenitor of all the insect tribes was a worm-like creature,

with a soft and fleshy body, a few jointed legs, and
the general appearance of a grub or caterpillar.
To this very ancient and somewhat shadowy
ancestor the larvæ of the higher insects still more
or less revert in their earlier stages; and we may
believe that many insects so reverted during many
generations. But in process of time the primitive

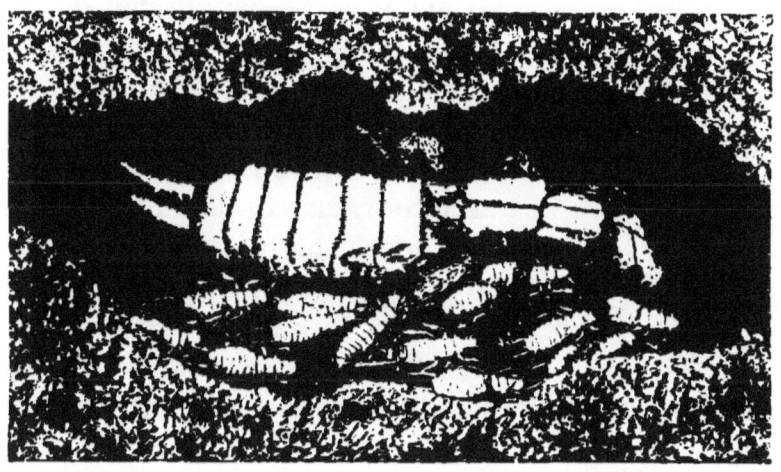

NO. 18.—THE MOTHER EARWIG AND HER BROOD OF CHICKS.

type developed into a wingless, six-legged form,
like that in No. 19—a form which you can see at
once marks a comparatively great advance upon
the old, worm-like progenitor. This animal, you
can note, has six good legs to run about with, and
is already provided with a well-marked head, and
with the three body-rings and the long tail or
abdomen so characteristic to the last of all higher
insects. Its segments have been specialised. From

such a type, it is probable the earwigs and their allies were developed by natural selection. But to this day every earwig begins life in a shape which closely resembles that of his first six-legged

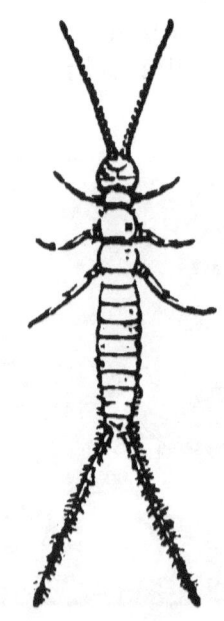

ancestor, and only gradually acquires his wings and other distinctively earwig-like features.

If you wonder how an animal so small as an earwig can do all the damage it undoubtedly does in gardens, a glance at No. 20 will explain the mystery. You will see from this sketch that the mouth-organs of the little beast are admirably adapted for destroying the petals of your choicest flowers. Nature has provided the earwig with a beautiful series of instruments for cutting holes in leaves and fruits. The figure in No. 20 is the lower part of the mouth, and is covered when at rest by the upper part, which is here placed below it. *M* are the mandibles or cutting jaws; they are formidable implements employed to saw holes in leaves, petals, or seed-capsules; while *C* is the *clypeus* or shield—in other words, the upper lip, which acts as a patent protector for the whole delicate apparatus. *AS* are the antennæ sockets, the feelers themselves having been removed for

NO. 19.—CAMPODEA, A PRIMITIVE WINGLESS INSECT.

After Sir John Lubbock.

the purposes of this sketch. The other parts of
the mechanism, I regret to say, can only be described in painfully technical language ; but as I
am generally sparing in my use of technicalities,

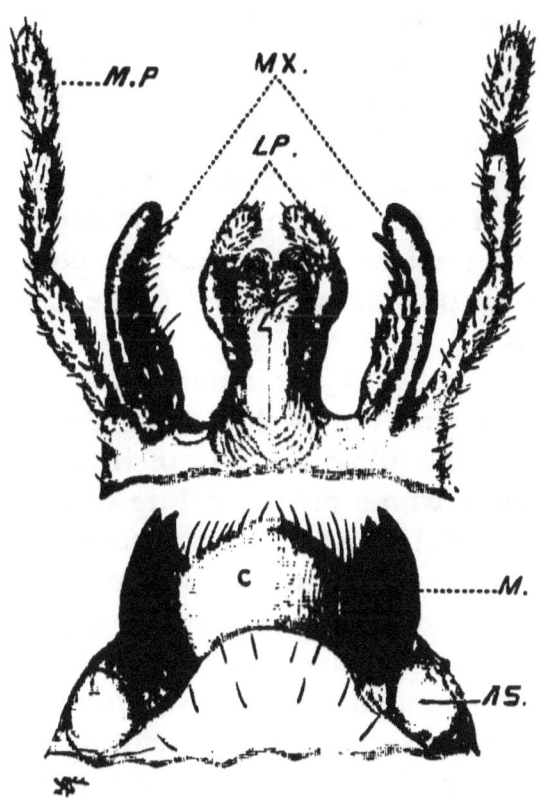

NO. 20.—THE EARWIG'S MOUTH, DISSECTED.

I trust I may be forgiven this solitary slip on the
ground of previous good conduct. *L* is the *labium*
or lower lip, which closes the mouth from below
when it is not in action. *LP* are the *labial palpi,*

K

used in manipulating the morsel as it is being eaten. *MX* are the *maxillæ*, or true jaws, employed in masticating the food, and answering in their functions pretty closely to the teeth of higher animals. Last of all, *MP* are the *maxillary palpi*, chiefly used like a pair of forks in holding the food, and, perhaps, also in deciding whether it is fit for eating. From this brief description, it will be immediately obvious to you that feeding with the earwig is a solemn and very complicated process. It is carried on by a number of distinct organs and implements, the exact purposes of each of which are only known at full to the insect which uses them.

I should add that the antennæ or feelers (not included in this last sketch, but conspicuous in all the previous illustrations) are in all likelihood sense-organs, whose precise nature has never been altogether established. Some naturalists believe that they are used as organs of smell ; others that they are combined organs of touch and guidance ; yet others, that they are the seat of a "sixth sense" unknown to humanity. However this may be, it is at least certain that they are useful as a means of communication between the insect himself and his mate, his young, his friends, and his acquaintances. Earwigs clearly feel their way, to a great extent, by the aid of the antennæ, and also recognise through them their visitors and family. They use them, too, in caressing or fondling their mates and their children. It is known that the antennæ are provided with numerous nerve-terminals, as is always the case with organs of the senses ; and I believe myself that, by their means, all insects of the same species are able to communicate more or less with one another by established signals. Perhaps the antennæ emit peculiar perfumes, which are recognised in turn by those of the friend or mate ; perhaps it is by touches and strokes that the insects transmit their ideas to one another. But that they do transmit ideas, nobody who has watched them closely ever doubts for a moment, and many naturalists even use the word "talking" of the parleys which ants and other insects carry on with their feelers.

It may be thought that an earwig's life, like a policeman's, "is not a happy one." This I hold to be an error. The earwig loves damp and darkness, it is true, but he flies at night in the beautiful twilight or by the soft rays of the moon, while his days are solaced by the companionship of his mate and his chosen comrades, for they are gregarious creatures. The mother tends her young with the assiduity of a hen sitting on her chickens, and food being abundant and cheap, life runs, as a rule, fairly smoothly with the earwig.

VII

THE FIRST PAPER-MAKER

THE civilised world could hardly get on nowadays without paper ; yet paper-making is, humanly speaking, a very recent invention. It dates, at furthest, back to the ancient Egyptians. "Humanly speaking," I say, not without a set purpose ; because man was anticipated as a paper-maker by many millions of years ; long before a human foot trod the earth, there is reason to suppose that ancestral wasps were manufacturing paper, almost as they manufacture it for their nests to-day, among the subtropical vegetation of an older and warmer Europe. And the wasp is so clever and so many-sided a creature, that to consider him (or more accurately her) in every aspect of life within the space of a few pages would be practically impossible. So it is mainly as a paper-manufacturer and a consumer of paper that I propose to regard our slim-waisted friend in this chapter.

It is usual in human language to admit, as the Latin Grammar ungallantly puts it, that "the masculine is worthier than the feminine, the feminine than the neuter." Among wasps, however, the

opposite principle is so clearly true—the queen or
female is so much more important a person in the
complex community, and so much more in evidence than the drone or male—that I shall offer
no apology here for setting her history before you
first, and giving it precedence over that of her
vastly inferior husband. *Place aux dames* is in this
instance no question of mere external chivalrous
courtesy ; it expresses the simple truth of nature,
that, in wasp life, the grey mare is the better
horse, and bears acknowledged rule in her own
city household. Not only so, but painful as it
may sound to my men readers, and insulting to
our boasted masculine superiority, the neuter in
this case ranks second to the feminine ; for the
worker wasps, which are practically sexless, being
abortive females, are far more valuable members
of the community than their almost useless fathers
and brothers. I call them neuter, because they
are so to all intents and purposes : though for
some unknown reason that seemingly harmless
word acts upon most entomologists like a red rag
on the proverbial bull. They will allow you to
describe the abortive female as a worker only.

In No. 1, therefore, I give an illustration of a
queen wasp ; together with figures of her husband
and of her unmarriageable daughter. The queen
or mother wasp is much the largest of the three ;
and you will understand that she needs to be so,
when you come to learn how much she has to do,
how many eggs she has to lay ; and how, unaided,
this brave foundress of a family not only builds a city

and peoples it with thousands of citizens, but also feeds and tends it with her own overworked mouth—I cannot honestly say her hands — till her maiden daughters are of age to help her. Women's-rights women may be proud of the example thus set them. Nature nowhere presents us, indeed, with a finer specimen of feminine industry and maternal devotion to duty than in the case of these courageous and pugnacious insects.

Male

Queen.

Worker.

NO. 1.—FAMILY PORTRAITS OF THE WASPS

But I will not now enlarge upon the features of these three faithful portraits, " expressed after the life," as Elizabethan writers put it, because as we proceed I shall have to call attention in greater detail to the meaning of the various parts of the body. It must suffice for the moment to direct your notice here to that very familiar portion of the wasp's anatomy, the sting, or ovipositor, possessed by the females, both perfect and imperfect—queens or workers—but not by those defenceless creatures, the males. The nature of the sting (so far as it is not already well known to most of us by pungent experience) I will enter into later ; it must suffice for the present to say that it is in essence an instrument for depositing the eggs, and that it is only incidentally turned into a weapon of offence or defence, and a means of stunning or paralysing the prey or food-insects.

The first thing to understand about a community of wasps is the way it originates. The story is a strange one. When the first frosts set in, almost all the wasps in temperate countries die off to a worker from the effects of cold. The chill winds nip them. For a few days in autumn you may often notice the last straggling survivors crawling feebly about, very uncomfortable and numb from the cold, and with their temper somewhat soured by the consciousness of their own exceeding weakness. In this irritable condition, feeling their latter end draw nigh, they are given to using their stings with waspish virulence on the smallest provocation ; they move about half-dazed on the damp ground, or lie torpid in their nests till death overtakes them. Of the whole populous city which hummed with life and business but a few weeks earlier, no more than two or three survivors at the outside struggle somehow through the winter, to carry on the race of wasps to succeeding generations. The colder the season, the fewer the stragglers who live it out ; in open winters, on the contrary, a fair number doze it through, to become the foundresses of correspondingly numerous colonies.

And who are these survivors ? Not the lordly and idle drones ; not even the industrious neuters or workers ; but the perfect females or queens, the teeming mothers to be of the coming communities. Look at the royal lady figured in No. 1. As autumn approaches, this vigorous young queen weds one of the males from her native nest. But shortly afterwards, he and all the workers of his city fall victims at once to the frosts of October. They perish like Nineveh. The queen, however, bearing all the hopes of the race, cannot afford to fling away her precious life so carelessly. That is not the way of queens. She seeks out some sheltered spot among dry moss, or in the crannies of the earth—a sandy soil preferred—where she may hibernate safely. There, if she has luck, she passes the winter, dormant, without serious mishap. Of course, snow and frost destroy not a few such solitary hermits ; a heavy rain may drown her ; a bird may discover her chosen retreat ; a passing animal may crush her. But in favourable circumstances, a certain number of queens do manage to struggle safely through the colder months ; and the wasp-supply of the next season mainly depends upon the proportion of such lucky ladies that escape in the end all winter dangers. Each queen that lives through the hard times becomes in spring the foundress of a separate colony ; and it is on this account that farmers and fruit-growers often

pay a small reward for every queen wasp killed
early in the spring. A single mother wasp destroyed in May is equivalent to a whole nest
destroyed in July or August.

As soon as warmer weather sets in, the dormant
queen awakes, shakes off dull sloth, and forgets
her long torpor. With a toss and a shake, she
crawls out into the sunshine, which soon revives
her. Then she creeps up a blade of grass, spreads
her wings, and flies off. Her first care is naturally
breakfast ; and as she has eaten nothing for five
months, her hunger is no doubt justifiable. As
soon, however, as she has satisfied the most pressing
wants of her own nature, maternal instinct goads
her on to provide at once for her unborn family.
She seeks a site for her nest, her future city.
How she builds it, and of what materials, I will
tell you in greater detail hereafter ; for the moment,
I want you to understand the magnitude of the
task this female Columbus sets herself—Columbus,
Cornelia, and Cæsar in one—the task not only of
building a Carthage, but also of peopling it. She
has no hands to speak of but her mouth, which
acts at once as mouth, and hands, and tools, and
factory, and stands her in good stead in her carpen-
tering and masonry. She does everything with her
mouth ; and therefore, of course, she has a mouth
which has grown gradually adapted for doing
everything. The monkey used his thumb till he
made a hand of it ; the elephant his trunk till he
could pick up a needle. Use brings structure ;
by dint of using her mouth so much, the wasp has

acquired both organs fit for her, and dexterity in employing them.

The first point she has now to consider is the placing of her nest. In this she is guided partly by that inherited experience which we describe (somewhat foolishly) as instinct, and partly by her own individual intelligence. Different races of wasps prefer different situations: some of them burrow underground; others hang their houses in the branches of trees; others again seek some dry and hollow trunk. But personal taste has also much to do with it; thus the common English wasp sometimes builds underground, but sometimes takes advantage of the dry space under the eaves of houses. All that is needed is shelter, especially from rain; wherever the wasp finds a site that pleases her, there she founds her family.

Let us imagine, then, that she has lighted on a suitable hole in the earth—a hole produced by accident, or by some dead mole or mouse or rabbit; she occupies it at once, and begins by her own labour to enlarge and adapt it to her private requirements. As soon as she has made it as big as she thinks necessary, she sets to work to collect materials for building the city. She flies abroad, and with her saw-like jaws rasps away at a paling or other exposed piece of wood till she has collected a fair amount of finely powdered fibrous matter. I will show you later on the admirable machine with which she scrapes and pulps the fragments of wood-fibre. Having gathered a sufficient quantity of this raw material to begin manufacturing, she proceeds to work it up with her various jaws and a secretion from her mouth into a sort of coarse brown paper ; the stickiness of the secretion gums the tiny fragments of wood together into a thin layer. Then she lays down the floor of her nest, and proceeds to raise upon it a stout column or foot-stalk of papery matter, sufficiently strong to support the first two or three layers of cells. She never builds on the ground, but begins her nest at the top of the supporting column. The cells are exclusively intended for the reception of eggs and the breeding of grubs, not (as is the case with bees) for the storing of honey. We must remember, however, that the original use of all cells was that of rearing the young ; the more advanced bees, who are the civilised type of their kind, make more cells than they need for strictly nursery purposes, and then employ some of them as convenient honey jars. The consequence is that beehives survive intact from season to season (unless killed off artificially), while the less prudent wasps die wholesale by cityfuls at the end of each summer.

Having thus supplied a foundation for her topsy-turvy city, our wasp-queen proceeds in due course to build it. At the top of the original column, or foot-stalk, she constructs her earliest cells, the nurseries for her three first-born grubs. They are not built upward, however, above the foot-stalk, but downward, with the open mouth below, hanging like a bell. Each is short and shallow, about a tenth of an inch in depth to begin with, and more like a cup, or even a saucer, than a cell at this early stage. The Natural History Museum at South Kensington possesses some admirable examples of such nests, in various degrees of growth ; and my fellow-worker, Mr. Enock, has obtained the kind permission of the authorities at

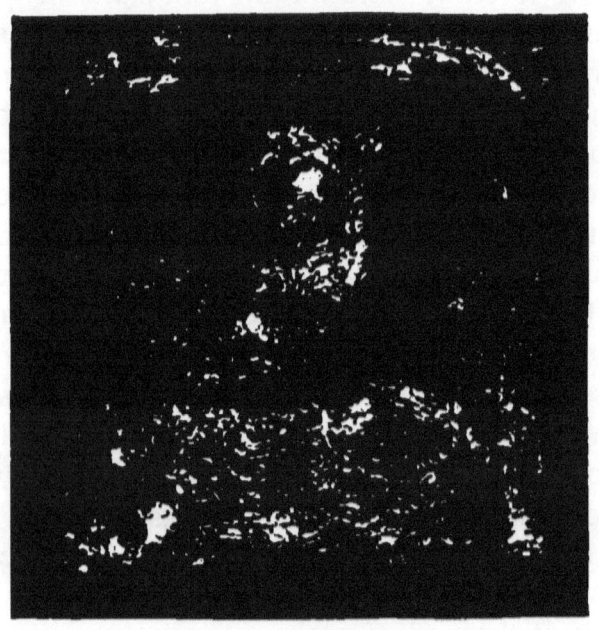

NO. 2.—THE CITY, TWO DAYS OLD.

the Museum to photograph the cases which contain them, for the purposes of these articles. They represent the progress of the queen-wasp's work at two, five, and fifteen days respectively (Nos. 2, 3, and 4), and thus admirably illustrate the incredible rapidity with which, alone and unaided, she builds and populates this one-mother city.

As soon as the first cells are formed in their early shallow shape, the busy mother, sallying forth once more in search of wood or fibre, proceeds to make more paper-pulp, and construct an umbrella-shaped covering above the three saucers. In each of the three she lays an egg ; and then, leaving the eggs to

NO. 3.—THE CITY, FIVE DAYS OLD.

hatch out quietly by themselves into larvæ, she goes on cutting—not bread and butter, like Charlotte in Thackeray's song—but more wood-fibre to make more cells and more coverings. These new cells she hangs up beside the original three, and lays an egg in each as soon as it is completed. But a mother's work is never finished ; and surely there was never a mother so hardly tasked as the royal wasp foundress. By the time she has built and stocked a few more cells, the three eggs first laid have duly hatched out, and now she must begin to look after the little grubs or larvæ. I have not illustrated this earliest stage of wasp-life, the

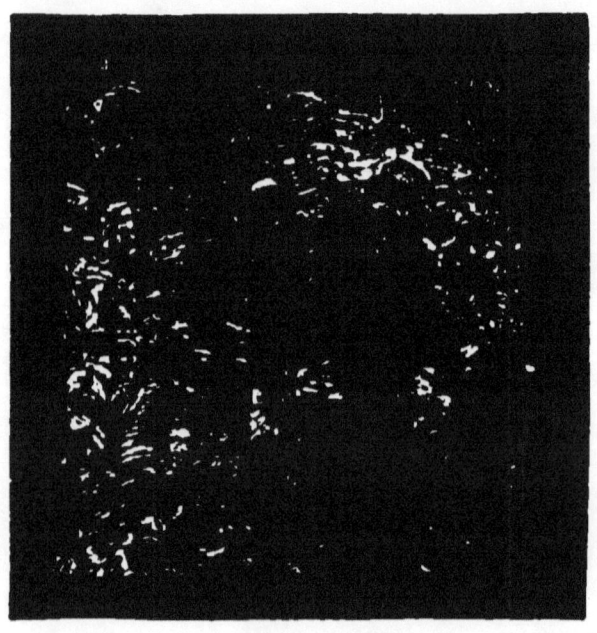

NO. 4.—THE CITY, FIFTEEN DAYS OLD.

grubby or nursery period, because everybody knows it well in real life. Now, as the grubs hatch out, they require to be fed, and the poor, overworked mother has henceforth not only to find food for herself, and paper to build more cells, but also to feed her helpless, worm-like offspring. There they lie in their cradles, head downward, crying always for provender, like the daughters of the horse-leech. Forgive her, therefore, if her temper is sometimes short, and if she resents intrusion upon the strawberry she is carting away to feed her young family by a hasty sting, administered, perhaps, with rather more asperity than a lady should display under trying circumstances. Some of my readers are mothers themselves, and can feel for her. Nor is even this all. The grubs of wasps grow fast—in itself a testimonial to the constant care with which a devoted mother feeds and tends them : and even as they grow the poor queen (a queen but in name, and more like a maid-of-all-work in reality) has continually to raise the cell-wall around them. What looked at first like shallow cups, thus grow at last into deep, hollow cells, the walls being raised from time to time by the addition of papery matter, with the growth of the inmates. In this first or foundation-comb—the nucleus and original avenue of the nascent city—the walls are never carried higher than the height of the larva that inhabits them. As the grub grows, the mother adds daily a course or layer of paper, till the larva reaches its final size, a fat, full grub, ready to undergo its marvellous metamorphosis. Then at last it begins to do some work on its own account : it spins a silky, or cottony, web, with which it covers over the mouth or opening of the cell ; though even here you must remember it derives the material from its own body, and therefore

ultimately from food supplied it by the mother.
How one wasp can ever do so much in so short
a time is a marvel to all who have once watched
the process.

While the baby wasps remain swaddled in their
cradle cells, their food consists in part of honey,
which the careful mother distributes to them impartially, turn about, and in part of succulent
fruits, such as the pulp of pears or peaches.   The
honey our housekeeper either gathers for herself
or else steals from bees, for truth compels me to
admit that she is as dishonest as she is industrious ;
but on the whole, she collects more than she robs,
for many flowers lay themselves out especially for
wasps, and are adapted only for fertilisation by
these special visitants.   Such specialised wasp-
flowers have usually small helmet-shaped blossoms,
exactly fitted to the head of the wasp, as you see
it in Mr. Enock's illustrations ; and they are for
the most part somewhat livid and dead - meaty
in hue.   Common scrophularia, or fig-wort, is
a good example of a plant that thus lays itself
out to encourage the visits of wasps ; it has small
lurid-red flowers, just the shape and size of the
wasp's head, and its stamens and style are so
arranged that when the wasp rifles the honey at
the base of the helmet, she cannot fail to brush off
the pollen from one blossom on to the sensitive
surface of the next.   Moreover, the scrophularia
comes into bloom at the exact time of year
when the baby wasps require its honey ; and you
can never watch a scrophularia plant for three

minutes together without seeing at least two or three wasps busily engaged in gathering its nectar. Herb and insect have learned to accommodate one another; by mutual adaptation they have fitted each part of each to each in the most marvellous detail.

It is a peculiarity of the wasps, however, that they are fairly omnivorous. Most of their cousins, like the bees, have mouths adapted to honey-sucking alone—mere tubes or suction-pumps, incapable of biting through any hard substance. But the wasp, with her hungry large family to keep, has to be less particular about the nature of her food; she cannot afford to depend upon honey only. Not only does she suck nectar; she bites holes in fruits, as we know to our cost in our gardens, to dig out the pulp; and she has a perfect genius for selecting the softest and sunniest side of an apricot or a nectarine. She is not a strict vegetarian, either; all is fish that comes to her net: she will help herself to meat or any other animal matter she can find, and will feed her uncomplaining grubs upon raw and bleeding tissue. Nay, more, she catches flies and other insects as they flit in the sunshine, saws off their wings with her sharp jaws, and carries them off alive, but incapable of struggling, to feed her own ever-increasing household.

By-and-by the first grubs, which covered themselves in with silk in order to undergo their pupa or chrysalis stage, develop their wings under cover, and emerge from their cases as full-grown workers. These workers, whose portrait you will find on a

L

previous page, are partially developed females, being unable to lay eggs. But in all other respects they inherit the habits or instincts of their estimable mother; and no sooner are they fairly hatched out of the pupa-case, where they underwent their rapid metamorphosis, than they set to work, like dutiful daughters, to assist mamma in the management of the city. Like the imagined world of Tennyson's " Princess," no male can enter. If ever there was a woman-ruled republic in the world, such as Aristophanes feigned, it is a wasp's nest. The workers fall to at " tidying up " at once; they put the house in order; they go out and gather paper; they help their mother to build new cells; and they assist in feeding and tending the still-increasing nursery. The first comb formed, you will remember, was at the top of the foundation column or footstalk; the newer combs are built below this in rows, each opening downward, so that the compound house or series of flats is planned on the exactly opposite system from our own—the top storeys being erected first, and the lower ones afterward, each storey having its floor above and its entrance at the bottom. At the same time, the umbrella-shaped covering is continued downward as an outer wall to protect the combs, until finally the nest grows to be a roughly round or egg-shaped body, entirely enclosed in a shell or outer wall of paper, and with only a single gateway at the bottom, by which the busy workers go in and out of their city.

The nest of the tree-wasp, which we have also

been kindly permitted to photograph from the specimens at the Natural History Museum (Nos. 5 and 6), exhibits this final stage of the compound home.

By the time the workers have become tolerably numerous in the growing nest, the busy mother and queen begins to relax her external efforts, and confines herself more and more to the performance of her internal and domestic duties. She no longer goes out to make paper and collect food ; she gives herself up, like the queen bee, exclusively to the maternal business of egg-laying. You must remember that she is still the only perfect female in the wasp hive, and that every worker wasp the home contains is her own daughter. She is foundress, queen, and mother to that whole busy community of 4000 or 5000 souls. The longer the nest goes on, the greater is the number of workers produced, and the faster does the queen lay eggs in the new cells now built for her use by her attentive daughters. These in turn fly abroad everywhere in search of nectar, fruits, and meat, or gather honey-dew from the green-flies, or catch and sting to death other insects, or swoop down upon and carry off fat, juicy spiders ; all of which foodstuffs, save what they require for their own subsistence, they take home to the nest to feed the grubs, from which, in due time, will issue forth more workers. It is a wonderful world of women burghers.

As long as summer lasts, our queen lays eggs which produce nothing else than such neuter workers. As autumn comes on, however, and the future of the race must be provided for, she lays eggs which hatch out a brood of perfect females or queens like herself. It is probable that the same egg may develop either into a queen or a worker, and that the difference of type is due to

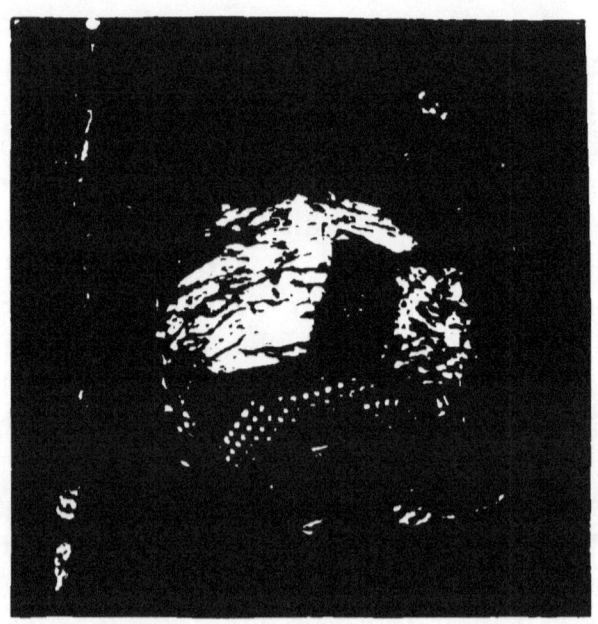

No. 5.—NEST OF TREE-WASP, WITH PAPER PARTLY REMOVED.

the nature of the food and training. A young grub fed on ordinary food in an ordinary cell becomes a neuter ; but a similar grub, fed on royal food and cradled in a larger cell, develops into a queen. As with ourselves, in fact, royalty is merely a matter of the surroundings.

Last of all, as the really cold weather begins to set in, the queen wasp lays some other eggs from which a small brood of males is finally developed. Nobody in the nest sets much store by these males : they are necessary evils, no more, so the wasps put up with them. It is humiliating to my sex, but I

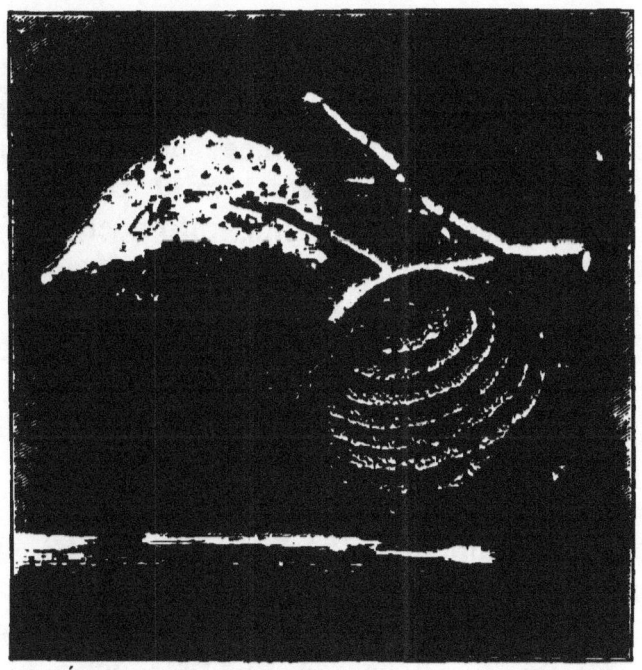

NO. 6.—NESTS OF TREE-WASP, EXTERIOR AND INTERIOR.

cannot avoid mentioning the fact, that the production of males seems even to be a direct result of chill and unfavourable conditions. The best food and the biggest cells produce fertile queens ; the second best food and smaller cells produce workers ; finally, the enfeeblement due to approaching winter

produces only drones or males. We cannot
resist the inference that the male is here the inferior creature. These facts, I regret to say, are
also not without parallels elsewhere. Among bees,
for instance, the eggs laid by very old, decrepit
queens, or by maimed and crippled queens, produce males only ; while among tadpoles, if well
fed, the majority become female frogs ; but if
starved, they become preponderantly male. So,
too, starved caterpillars produce only male butterflies, while the well-fed produce females. I know
this is the opposite of what most people imagine ; but then, science not infrequently finds
itself compelled to differ in opinion from most
people.

The drones, or males, are thus of as little account
in the nest of wasps as in the hive of bees. In
both, they only appear for a short time, and for
the definite purpose of becoming fathers to the
future generations. When they have fulfilled this
their solitary function, the hive, or the nest, cares
no more about them. The bees, as you know,
have a prudent and economical habit of stinging
them to death, so as not to waste good honey
on useless mouths through the winter. The wasps
act otherwise. They are not going to live through
the winter themselves, so they don't take the trouble
to execute their brothers : they merely turn the
young queens and males loose, and then leave the
successful suitors to be killed by the first frost
without further consideration.

And now comes the most curious part of all

this strange, eventful history. We do not love wasps ; yet so sad a catastrophe as the end of the nest cannot fail to affect the imagination. As soon as the young queens and males have quitted the combs, the whole bustling city, till now so busy, seems to lose heart at once and to realise that it is doomed to speedy extinction. Winter is coming on, when no worker wasp can live. So the community proceeds with one accord to commit communal suicide. The workers, who till now have tended the young grubs with sisterly care, drag the remaining larvæ ruthlessly from their cells, as if conscious that they can never rear this last brood, and carry them in their mouths and legs outside the nest. There they take them to some distance from the door, and then drop them on the ground to die, as if to put them out of their misery. As for the workers themselves, they return to the nest and starve to death or die of cold ; or else they crawl about aimlessly outside in a distracted way till the end overtakes them.

There is something really pathetic in this sudden and meaningless downfall of a whole vast cityful ; something strange and weird in this constantly repeated effort to build up and people a great community, only to see it fall to pieces hopelessly and helplessly at the first touch of winter. Yet how does it differ, after all, from our human empires, save in the matter of duration ? We raise them with infinite pains only to see them fall apart, like Rome or Babylon.

So, by the time the dead of winter comes, both males and workers are cleared off the stage; and universal waspdom is only represented by a few stray fertilised females, who carry the embodied hopes of so many dead and ruined cities.

And now that I have traced the history of the

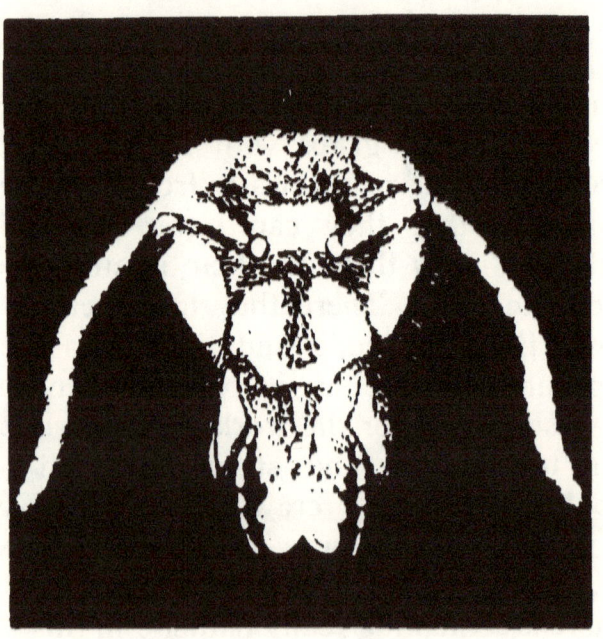

No. 7. —HEAD OF QUEEN WASP, MOUTH WIDE OPEN; FRONT VIEW.

commune from its rise to its fall, I must say a few words in brief detail about the individual wasps which make up its members.

And first of all as to the wasp's head. You will have gathered from what I have said that the head of the insect is practically by far its most important portion. All the work we do with our hands, the wasp does with its complicated mouth-organs. And the wasp's head is such a wonderful mechanism, that some little study of the accompanying illustrations, though they may not at first sight look very attractive, will amply repay you. I will

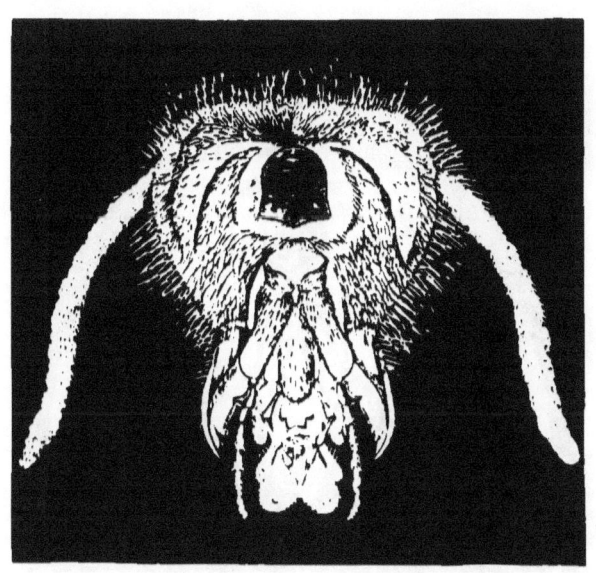

NO. 8.—THE SAME HEAD, MOUTH WIDE OPEN : BACK VIEW (DECAPITATED).

try to explain the uses of each part with as little as possible of scientific technicalities.

In No. 7 you get the head of a queen wasp, seen full face in front, with the mouth-organs open. The three little knobs in the centre up above are the simple eyes or eyelets (*ocelli*, if you prefer a Latin word, which sounds much more learned).

The large kidney-shaped bodies on either side of the head (there seen as interrupted by the antennæ or feelers) are the compound eyes, each of which consists of innumerable tiny lenses, giving the wasp that possesses them a very acute sense of vision. We do not know exactly what is the difference in use between the simple eyes and the

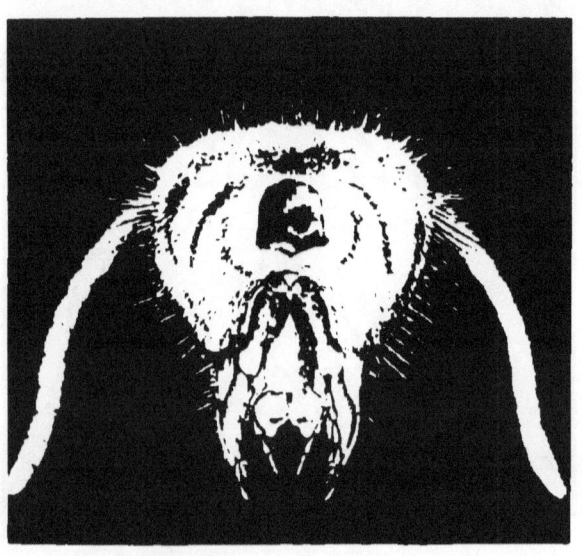

NO. 9.—THE MOUTH CLOSING: TONGUE WITHDRAWN: BACK VIEW

compound ones; but either sort has doubtless its own special part to play in this complex personality. The antennæ, or feelers, again, with their many joints and their ball-and-socket base, are beautiful and wonderful objects. The various parts of the mouth are here seen open; conspicuous among them are the great saw-like outer jaws, used for scraping wood and manufacturing paper ; the long, narrow shield ; the broad tongue ; and the delicately jointed palps, or finger-like feeders.   Notice how some of these organs are suitable for cutting and rasping, while others lend themselves to the most dainty and delicate manipulation.

No. 8 shows us the same head, decapitated, and

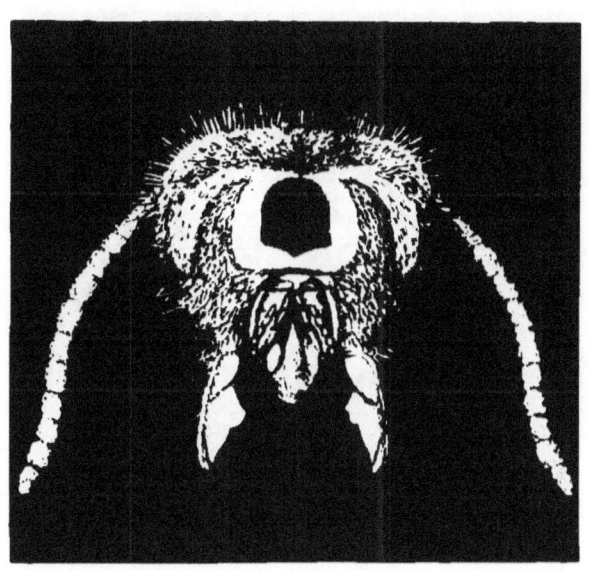

NO. 10.—MOUTH ALMOST CLOSED : ATTITUDE FOR SCRAPING WOOD : BACK VIEW.

seen from behind.   The shield-like space in the very middle represents the point of decapitation—the cut neck, if I may use frankly human language. Below is the hollow or receptacle into which all the organs can be withdrawn when not in use, and packed away like surgical knives and lancets in an instrument case.   Observe in the sequel how neatly and completely this can be done: how each has its groove in the marvellous economy of nature.

In No. 9 you see the organs closing (also a back view), the tongue having been now drawn in, while the saw-like jaws and the delicate feeling palps are still exposed and ready for working. No. 8, on the contrary, is the feeling attitude.

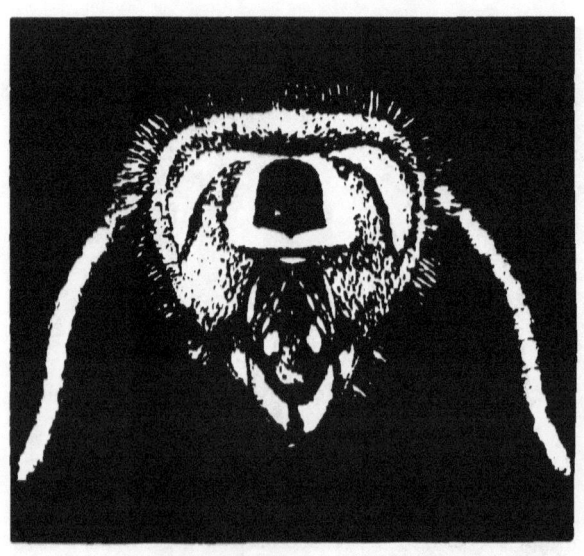

NO. 11.—MOUTH QUITE CLOSED: ATTITUDE FOR SCRAPING WOOD END OF ONE MOVEMENT.

In No. 10 (another back view), the palps have been turned back into their special groove, and the saw-like jaws are seen free for working. This is the attitude in which the wasp attacks a park paling, in order to scrape off wood-fibre for the manufacture of paper. Here, as you see, the jaws are open. In No. 11 they are closed, at the end

of a scrape.   These two last attitudes are, of course,
alternate.   One shows the jaws opened, the other
closed, as they look at the beginning and end of
each forward and backward movement.   You will
notice also that, as usual, the insect's jaws work
sideways, not up and down like those of man and
other higher animals.   If you examine closely this
series of wasp's heads in different postures, you
will see how well the various parts are adapted,
not only for rasping and manu-
facturing paper, but also for
the more delicate work of wall
and cell building.

Almost as interesting as the
head are the wings of wasps,
of which there are four, as in
most other insects.   But they
have this curious peculiarity :
the two front wings have a
crease down the middle, so that
they can be folded up lengthwise, like two segments or rays
of a fan, and thus occupy only

NO. 12.—QUEEN WITH
FOLDED WINGS, AND
ONE WING TO SHOW
FOLDING.

half the space on the body that they would otherwise do.   It is this odd device that makes the
transparent and gauzy wings so relatively inconspicuous when the insect is at rest, and the same
cause contributes also to the display of the handsome black-and-yellow-striped body.   No. 12 shows
us a queen with her wings folded : below is one
upper or front wing, folded over on itself, and then
laid across the under wing.   No. 13 introduces us

to a more characteristic feature, common to wasps with the whole bee family.

All these cousins possess by common descent the usual four wings of well-regulated insects. But it so happens that the habits of the race make strong and certain flight more practically important for them than the mere power of aërial

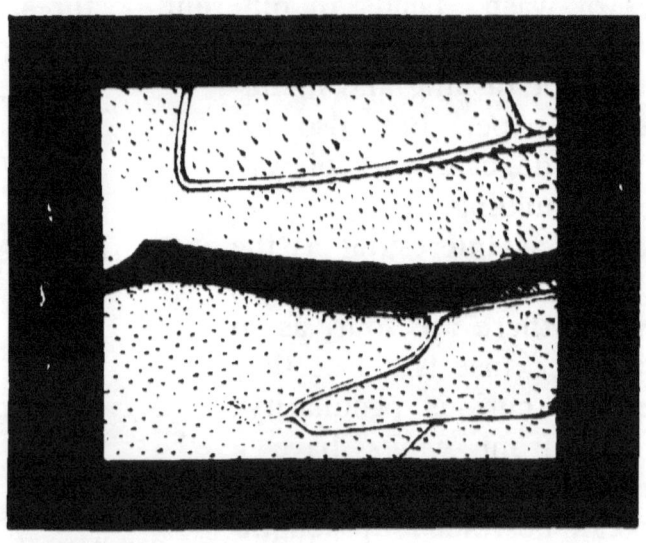

No. 13.   PART OF TWO WINGS, WITH HOOKS AND GROOVES.

coquetting and pirouetting possessed by the far less business-like butterflies.   Your wasp and your bee are women of business.   They have therefore found it pay them to develop a mechanism by which the two wings on either side can be firmly locked together, so as to act like a single pinion. No. 13 very well illustrates this admirable plan for fastening the fore and hind wings together.

On top you see the back portion of the front wing, with a curved groove on its inner edge. Below, you get the front portion of the hinder wing, with a series of little hooks, microscopic, yet exquisitely moulded, which catch into the groove on the opposite portion.   When thus hooked together, the two wings on the right act exactly like one.   So do the two on the left.   But they can be un-hooked and folded back on the body at the will of the insect. To either side of No. 13 you will notice sections of the two wings, which will help you to under-stand the nature of the mechan-ism.   On the right, the wings are seen hooked together ; on the left, they are caught just in the act of unhooking.

Last of all, and most important of all to ordinary humanity, we come to the sting, with its append-age the poison-bag.   It is well represented in No. 14.   The main object of the sting, and its ori-ginal function by descent, is that of laying eggs ; it is merely the ovipositor.   But besides the grooved sheath or egg-layer (marked S in the illustration) and the two very sharp lances or darts (marked D) which pierce the flesh of the enemy, it is provided with a gland which secretes that most unpleasant body, formic acid ; and when the wasp has cause

NO. 14.—POISON BAG, SHEATH, DARTS, AND PALPI.

to be annoyed, she throws the sting rapidly into the animal that annoys her, and injects the fluid with the formic acid in it. In No. 15 the darts are shown still more highly magnified. In the queen wasp, the sting is used both for laying eggs and as a weapon of offence ; but in the workers, which cannot lay eggs, it is entirely devoted to the work of fighting.

NO. 16. — WASP'S BRUSH AND COMB, FOR CLEANING ANTENNÆ.

Two other little peculiarities of the wasp, however, deserve a final word of recognition. One

NO. 15.—DARTS MAGNIFIED 300 DIAMETERS.

of these is the elaborate brush-and-comb apparatus or antennæ-cleaner, drawn in a very enlarged view in No. 16. Whatever the sense may be which the antennæ serve, we may at least be certain that it is one of great import-

ance to the insect ; and both wasps and bees have
therefore elaborate brushes for keeping these valuable organs clean and neat and in working order.
They always remind me of the brushes I use myself
for cleaning the type in my typewriting machine.
The antennæ-brush of the wasp is fixed on one
of her legs ; its precise situation on the leg as
a whole is shown in the little upper diagram ;
its detail and various parts are
further enlarged below. To the
left is the coarse or large-tooth
comb ; to the right is the brush ;
and above the brush, connected
with the handle by an exceedingly thin and filmy membrane,
is the fine-tooth comb, used for
removing very small impurities.
With this the wasp cleans her
precious feelers much as you
may have seen flies clean their
wings when they have fallen
in a jam-pot ; only the wasp's
mechanism is much more beautiful and perfect.

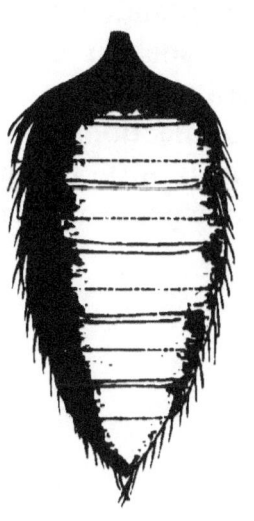

NO. 17.—TUCKS IN THE
SEGMENTS.

Almost equally interesting with the brush and
comb are the series of tucks in the wasp's body
or abdomen, delineated in No. 17. By means of
these extraordinarily flexible rings, each held in
place or let loose by appropriate muscles, the
wasp can twist her body round so conveniently
that, no matter how carefully and gingerly you
hold her, she will manage to sting you. They

M

are models of plate-armour. They work upward, downward, and more or less sideways, so that they enable her to cock her body up or down, right or left, at will, with almost incredible flexibility.

Adequately to tell you all about the wasp, however, would require a very stout volume. I have said enough, I hope, to suggest to you that the wasp's history is quite as interesting as that of her over-lauded relation, the little busy bee. Indeed, I suspect it is only the utilitarian instinct of humanity that has caused so much attention to be paid to the domestic producer of honey, and so relatively little to that free and independent insect, the first paper-maker.

VIII

ABIDING CITIES

THE papery nests of wasps are purely temporary empires: the vespine race has "no abiding city here"; each summer sees the populous homes built afresh from the ground; each winter sees them unpeopled and demolished. But with ants, which are builders for time, things are quite otherwise. The communities of those clever and intelligent little creatures are tolerably permanent; they go on from year to year, and generation to generation, often for very long periods together. Lest I weary you unnecessarily by a long preamble, however, I shall present you with views of one such nest at once, outside and inside, in Nos. 1 and 2, in order that you may see without delay the curious method of their detailed construction.

The city whose external lineaments are shown you in the photograph reproduced in No. 1 is actually situated on St. George's Hill, near Weybridge, just ten feet away from the large Scotch fir whose trunk appears on the right of the illustration. It is only one among many various types of ants' nests built by different species. From outside, all you can see of it is a confused mass of dry pine-needles, arranged in a barrow-shaped hill or mound, some eight feet across at the base, and two

NO. I.—A WOOD ANTS' NEST, EXTERIOR.

feet high. But that is in reality only the outwork or top storey of the communal habitation. Beneath

it lies a second layer, six inches thick, composed
entirely of roots of heather and rootlets of fir-trees,
all carefully stripped clean of bark, and making a
dry foundation for the warm hillock of pine-needles.

NO. 2.—A WOOD ANTS' NEST, INTERIOR; EGGS, GRUBS, AND
COCOONS, WITH WORKERS ENGAGED IN TENDING THEM.

Below this woody layer, again, the ground is tunnelled to an unknown depth by long subterranean
galleries, driven right through a stratum of solid
sandstone. These inner galleries extend not only

beneath the hillock, but also all round it, for
wherever you step the soil treads soft, and gives
beneath your foot to a depth of six or eight inches.
This illustrative example is a city built by the common Wood Ant. I have had another just like it
—an insect London—under observation for three
or four years in a copse on a spur of Hind Head,
not far from my cottage.

In No. 2 Mr. Enock has represented for us, with
his usual skill, a very small section of such a city,
" all a-growing and a-blowing,"—all engaged in
the active exercise of its everyday functions. How
it came into being, and how it is ruled and peopled,
I will tell you a little later on ; for the present, I
want first to familiarise you with the general course
of its domestic economy in practical action. We
have here an interior view, with one wall removed,
of a tunnel or gallery, which runs through the soft
upper portion of the nest, composed of pine-
needles ; together with a small piece of the outer
surface. An ant, which has been out foraging for
food, approaches one of the mouths of the nest.
Beneath are three successive floors or stages of the
tunnel, with excavated chambers, each appropriated
to its own particular purpose. In the upper floor
of all, we see two groups of minute eggs awaiting
their hatching. These are the real eggs, not the
much larger things sold as "ants' eggs" for bird
food in London, which are really the pupæ. Four
of the eggs have just arrived at hatching point ;
therefore, one of the careful nurses who look after
them is seen just in the act of bundling them over

on to stage two, which is the floor here reserved
for the nursery of the hatched-out grubs or larvæ.
In this second stage you see a chamber with a
group of such grubs, all hungry and greedy, waiting for their nurses to bring them food from outside
the household.   Observe the obvious expectancy of
their attitude, with heads held up, like that of small
birds clamouring eagerly for food when their mother
approaches them with a worm or a caterpillar.
After feeding for some time in this legless, grub-
bish condition, the larva turns into a pupa, and
encloses itself in a cocoon.   One larva has just completed this happy transformation, and a watchful
nurse ant is therefore at this moment engaged in
carrying it tenderly a stage lower down to the floor
reserved for the chrysalis condition.   On the third
floor, below, you see a group of pupæ lying by in
the dark, and awaiting their development.   The
wall of one cocoon has here been removed, and
within you may catch a glimpse of the imprisoned
grub, now recently transformed into the adult ant
pattern.   Of course, the nest contains many hundreds of such tunnelled galleries, all teeming with
life, and all made up of several distinct chambers.

Now, how does such a nest begin to be ?   Well,
it starts from a queen, or perfect female, who sets
out with a few others to form a colony.   This
colony soon grows, but it is rather a republic than
an Amazon kingdom, like the hive of bees or the
nest of wasps.   It is composed of several perfect
females (instead of one queen), numerous imperfect females or workers, and a few males, who, as

is usual among social insects, are very unimportant and unconsidered creatures. The males and females are winged when they first emerge from their cocoons, and they use their wings for their marriage flight, which is a recognised institution among all insect socialists. But as soon as the perfect females have been safely wedded, their wings drop off; or, in cases where they do not fall of themselves, the insects themselves wriggle and pull them off with their legs in the most comic fashion. I have sometimes seen a dinner-table in Jamaica covered by a sudden irruption of female winged ants of tropical species, which insisted on immolating themselves in the soup and the wine (to the advantage of neither party), while others blackened the table-cloth, and devoted themselves to getting rid of their wings with unpleasant gyrations. As for the males, they are of no further use to the community, so they die at once. But the mass of the larvæ develop into imperfect females or workers, which are always wingless from the very first, and it is these that form the ordinary ants of the everyday observer. In many kinds there are also two types of neuters: the one type, workers proper, have rather large heads and moderate jaws —they are the foragers and builders of the community; the other type, soldiers, have still bigger heads and very powerful jaws—it is their task to fight in defence of their native city. Other differences of less importance will come out in the course of our subsequent explanation.

The winged ants have large and many-faceted

compound eyes, to aid them in their flight abroad ;
and they have also single eyelets or *ocelli*, as in the
case of the wasp, which seem to be useful to them
in finding the way over large areas, as the compound eyes are probably designed for nearer and
minuter vision. But the workers have always the
true eyes small, and often rudimentary ; while the
eyelets or *ocelli* are mostly wanting. To put it
plainly, they are almost blind. There can be very
little doubt that their principal organ of sense
resides in the antennæ, or feelers, which are probably used in part for smelling. Whatever may
be the perceptive function which these curious
appendages subserve, however, nobody who has
watched ants closely ever doubts that they are also
used as a means of intercommunication, almost
analogous to human language. Whenever two
ants of the same nest meet, they stop and parley
with one another by waving and crossing their
antennæ ; so obvious is it, that the information
thus conveyed makes one ant follow another towards a source of food, or other object of interest,
which the first ant has discovered, that the process is universally described by ant observers as
" talking."

In No. 3 we get an illustration of two workers
belonging to an English species known as the
Warrior Ant, from its predatory habits, engaged
in just such a profound confab together. They
are meditating war, and discussing a plan of cam-
paign with one another ; for the Warrior Ant is a
slave-making species. It is a large red kind, and it

makes raids against nests of the small yellow Turf Ant, a mild and docile race, large numbers of which it carries off to act as servants. But it does not steal fully-grown Turf Ants ; their habits are formed, and they would be useless for such a purpose. What the Warrior Ant wants is raw material which can be turned into thoroughly well-trained servants. So it merely kills the adult ants which strive to oppose its aggression, and contents itself

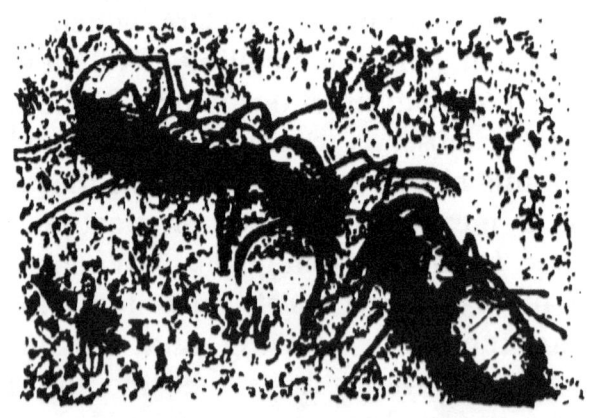

NO. 3.—A CONVERSATION : "LET'S GO SLAVE-HUNTING :"

with trundling home to its own nest the larvæ and pupæ of the Turf Ants which it has put to flight and vanquished. In process of time, these grubs and cocoons produce full-grown yellow workers, which, having never known freedom, can be taught by the Warrior Ants to act as nurses and house-maids, exactly as if they were living in their own proper city. I once saw in a garden in Algiers a great pitched battle going on between slave-makers and the family of the future slaves, in which the ground was strewn with the corpses of the vanquished. Not till the nest of the smaller ants was almost exterminated did they retire from the unequal contest, and allow the proud invader to carry off their brothers and sisters in their cocoons, asleep and unconscious.

The two ants figured in No. 3 are deliberating on the chances of such a cocoon-lifting expedition. The one to the right has been hunting for honey up the stems of vetches, and has fallen in by the way with a small nest of Turf Ants. Returning post-haste to her own home, big with this exciting intelligence, she encounters a comrade, to whom she communicates, in antennæ language, her belief that the Turf Ants she has discovered are not very numerous, and her conviction that they would fall an easy prey to a well-organised party of Warrior raiders. The two friends cross their antennæ as they talk, wave them mysteriously about, and evidently succeed in conveying their respective views on the situation to one another. After a short delay, both return, all agog, to the nest together, and rouse the guard with intelligence of plenty of pupæ ready to be plundered. At once the city hums, alive with bustle and preparation. Workers run to and fro and communicate orders from headquarters to one another. "There's a big slave-hunt on ; sister-fighter so-and-so has just brought news of a city of Turfites, quite near, and unprotected. The doors are open, and she noticed as she passed that the sentries looked most lax and indifferent. The whole place has apparently been demoralised by a recent marriage flight. Everybody in our nest is going to the war. Come along and help us!"

Forthwith they sally out, and make for the city of the despised yellow Turfites. They fall upon it unexpectedly, and kill the outer sentries. Then the battle begins in earnest. Half the Turfites rush out in battle array, and, banding themselves together, to make up for their individual small size, fall fiercely upon this or that isolated Warrior. Occasionally, by dint of mere numbers, they beat off the invader with heavy loss; but much more often, the large and strong-jawed Warriors win the day, and destroy to a worker the opposing forces. They crush their adversaries' heads with their vice-like mandibles. Meanwhile, within the nest, the other half of the workers—the division told off as special nurses—are otherwise employed in defending and protecting the rising generation. At the first alarm, at the first watchword passed with waving antennæ through the nest, "A Warrior host is attacking us!" they hurry to the chambers where the cocoons are stored, and bear them off in their mouths into the recesses of the nest, the lowest and most inaccessible of all the chambers. When at last the day is lost, the Warriors break in and steal all the pupæ they can lay their jaws upon; but many survive in the long, dark tunnels, with a few devoted workers still left to tend and teach them.

No. 4 shows us the final stage in such a slave-

hunt. The big red Warriors have won ; the little
yellow Turfites have been repulsed and defeated
with great slaughter. The victors are at present
engaged in carrying captured cocoons to their own
nests ; there the pupæ will hatch out shortly into
willing slaves, and, never having known any other
condition, will take it for granted that the natural
post for small yellow ants is to clean and forage
and catch food for big red ones.

Our own Warrior Ants are slave-holders which

NO. 4.—A SLAVE-HUNT ; CONQUERORS CARRYING OFF THE
COCOONS OF THE ENEMY.

still retain some power of working and acting for
themselves ; but there are other species in which
the " peculiar institution " has produced its usual
degrading result by rendering the slave-owner incapable and degenerate, a mere fighting do-nothing.
Among the Amazon ants, which are very confirmed
slave-makers, Sir John Lubbock found that a great
lady, left alone without slaves, in the presence of
food, did not even know how to feed herself ; she
was positively starving to death in the midst of

plenty. Then Sir John provided her with a single slave; instantly, the industrious little creature set to work to clean and arrange her mistress, and to offer her food. This is a striking illustration of the moral truth that slavery is at least as demoralising for the master as for his servant.

No. 5 introduces us to a passing phase in a combat of ants—a life-and-death conflict between

two single antagonists. Ants, indeed, are desperate fighters; the workers and perfect females have sometimes stings, like the bees and wasps; but in most species they fight by biting with their jaws, which are moulded into strong

NO. 5.—PAYING OFF OLD SCORES: A LIFE-AND-DEATH CONFLICT.

and vice-like nippers or pincers. Moreover, they have a gland which secretes the same poisonous material as that contained in the venom-bag of the sting among wasps and bees; and after the ant has made a hole with her jaw in her enemy's armour, she injects into it a little of this painful irritating acid, which kills small insects. During a battle, ants are all most reckless of their own lives; indeed, no ant seems ever to consider herself by compari-

son with the interests of the community at large.
The individual exists for the estate alone, and
sacrifices her life and happiness, automatically as
it were, on behalf of her city.

In No. 6 we see an illustration
of the great muscular strength
possessed by ants, especially in
their gripping jaws or mandibles.
Here, two comrades have got hold
of a dead and rigid prey, which
they are striving to carry off by
main force to the nest ; for ants
are omnivorous. They feed off
whatever turns up handy ; all is
fish that comes to their net—
they seem almost indifferent
whether what they dine off is
honey or honeydew, a worm or a
beetle, a dead
bird or a de-
parted lizard.
A few workers
will seize what-
ever edible
object they
happen to find,
and combine
to drag it

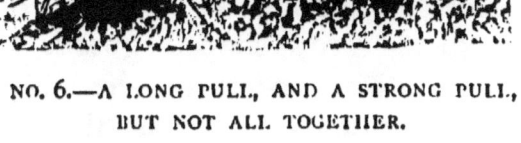

NO. 6.—A LONG PULL, AND A STRONG PULL,
BUT NOT ALL TOGETHER.

away, by pushing and pulling, to the underground
chambers. In this particular case the two ants
began by hauling together ; but the lower one,
giving one good tug with her jaws, has succeeded

in raising the whole carcass aloft, and hoisting up
her astonished neighbour into the air on top of it.
It is impossible to watch a nest of ants at work
for any length of time without being the spectator
of many such comic little episodes.

I implied above that ants are very fond of honey.
But plants by no means desire their attentions ;
because, being creeping creatures, guided mainly
by the sense of smell, they crawl up the stems of
one species after another, indiscriminately, and so
do no good in setting the seeds of any particular
kind of flower. To baffle them, accordingly, many
plants cover their stems with downward-pointing
hairs, which prove to the ants as impenetrable an
obstacle as tropical jungles to the human explorer ;
while other sorts set various traps like lobster-
pots on their stalks, to catch and imprison the
unwelcome visitors. But the wild vetches have
a still more curious and instructive habit, shared
by not a few other ingenious plants. They buy
off the intruders by an organised system of black-
mail. Below the flowers intended for fertilisation
by flying insects, which flit straight from one
blossom to another of the same kind, the vetches
put some arrow-shaped guards or stipules, so
arranged like barriers on the stem that a prying
ant cannot easily creep past them. In the centre
of each stipule, however, the plant produces a little
black gland, which secretes honey. This honey is
a bribe to the marauding ant ; the vetch puts it
there in order that the insect, finding its progress
toward the flower blocked, may just stop *en route*

and sip this pittance of nectar, leaving the richer and more valuable stock of honey in the actual blossom to be rifled by the bees which are the honoured guests and allies of the vetches. Nature is all full of such quaint plots and counterplots. One example occurs in a South American tree, so very remarkable that I cannot pass it by even in this hasty notice.

A certain ant, very common in Brazil, has the habit of cutting large round pieces out of the leaves of trees, which it then conveys to its nest for the purpose of growing fungi upon them—in human language, making tiny mushroom-beds. Now, this habit is naturally obnoxious to the trees, which produce the leaves for their own advantage, not for the sake of leaf-cutting ants which hack and rob them. To guard against the burglarious leaf-cutters, accordingly, one clever South American acacia has hit upon an excellent plan of defence. It produces curious hollow thorns ; while each leaflet has a gland at its base which secretes honey. Into these hollow thorns, colonies of a small and harmless ant migrate, and take up their abode there. They live off the honey at the base of the leaflets. They thus acquire a vested interest in the acacia tree, which is their home and territory ; and whenever the leaf-cutting ants attack the acacia, the little occupants of the thorns and owners of the honey-chambers pour out upon them in their thousands, and compel the invaders to beat a hasty retreat with heavy losses. Thus the cunning tree supplies its insect body-guard with board and lodging in return for efficient protection against the dreaded onslaught of the common enemy.

And now that I have succeeded, I hope, in interesting you a little in the habits of ants, I am going to tell you a few facts about their structure. That is my dodginess, you see; I knew if I began by giving you details of legs and body and segments, you would vote the whole thing dry; but

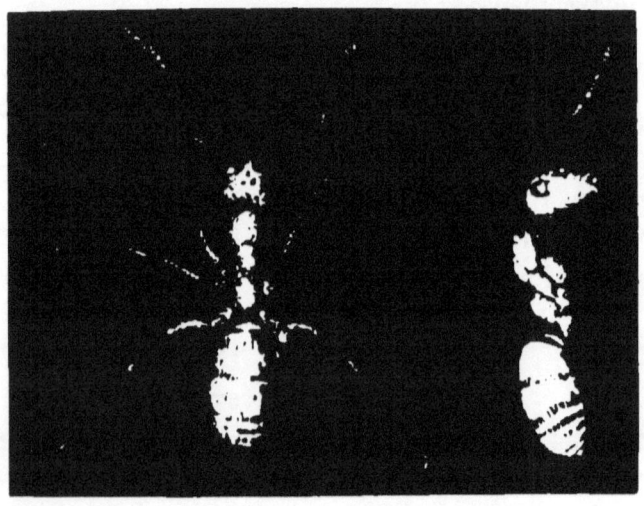

NO. 7.—THE GARDEN ANT PORTRAIT OF A WORKER.

now that you understand what sort of objects the ant wants to attain, you may be content to examine the organs she attains them with.

In No. 7 you have a portrait of the common Garden Ant of England, one of the most interesting creatures in the world to watch in action. This is a worker specimen; therefore, it has a very big head, with very powerful jaws; and when

you remember that ants work for the most part
with the head only, you will understand why that
portion needs to be the most muscular and powerful part of the body. A lobster has two very
strong claws in front, because those are his fighting and prey-catching organs ; the ant's jaws
just answer in function to the lobster's claws,
and to our hands and arms, and, therefore, they
are correspondingly big and muscular. Male and
female ants do not have to dig tunnels, to build
up chambers, to drag heavy weights back to the
nest ; therefore, they have smaller heads and
bigger eyes ; they are adapted only for flying
and for producing the younger generation. The
middle segments of the body, on the contrary,
are large and powerful in the males and females,
because they have to work the wings ; while in
the workers they are smaller, especially in one
segment, because the workers are wingless. The
legs, however, are fairly strong, since they need
to pull and to supply a firm footing when the
ant is tugging hard at some heavy object. But
between the part of the body which forms the
attachment for the six legs and the abdomen,
or "tail," there is a single characteristic segment,
or stalk, very thin and slender, which bears a
sort of scale, peculiar to the ant family. The
side view, with the legs removed, enables you
to note how admirably the ant is adapted for
turning in almost any direction, and explains
that extraordinary flexibility of body which you
must have noticed whenever you have watched

a troop of ants trying to drag a dead insect over a gravel path, and surmounting all obstacles with clumsy ingenuity. Ants, in short, are built for navvies; they are insect engineers, and they have acquired a form exactly adapted to their peculiar habits.

But why are the worker ants so nearly blind? That must surely be a disadvantage to them.

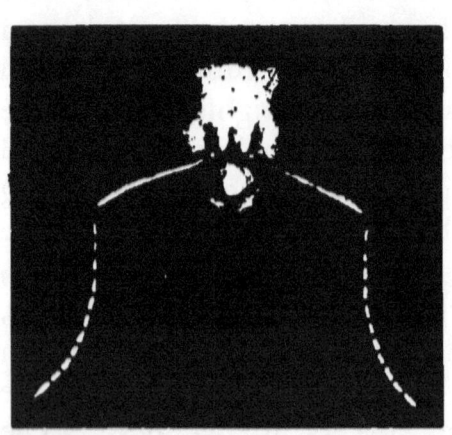

NO. 8.—HEAD OF GARDEN ANT, WITH EYES, ANTENNÆ, JAWS, AND FEELERS

Not a bit of it. Ants work mainly in dark underground passages, where the sense of sight would be of little use; and, moreover, like all hunting animals, they find smell more important as an indicator of food in the open than vision. The hound does not *look* for the fox—he sniffs and scents him. Now, whenever any sense is relatively unimportant, an economy may be effected by suppressing or curtailing it; the material that would otherwise go to making and repairing its organ is more profitably employed on some better work elsewhere. Ants are obviously descendants of flying ancestors, none of which were workers; and the flying males and females possess to this day the organs of sight necessary for their habits. But in the class of workers it has been found more useful, on the whole, to concentrate attention on smell and on strength of jaw than on sight and flight: the important point is that the worker ant should be able to find scattered foodstuffs, and should be strong enough to pull them back to the city. So in No. 8 you get a front view of the head of the common Garden Ant ; and you will see for your-self that its eyes, when compared with the nume-

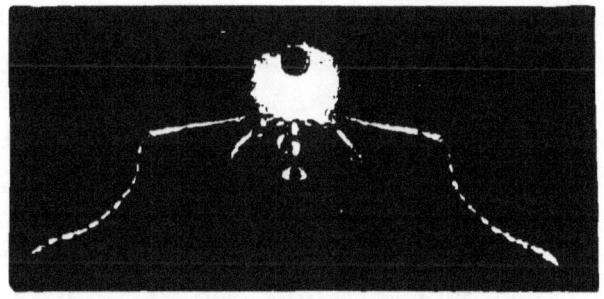

NO. 9.—BACK VIEW OF HEAD, WITH JAWS OPEN, AND ORGANS EXPANDED.

rous eyelets and large compound organs of the wasp, are relatively imperfect ; while its antennæ are large and fully developed appendages. They turn in a beautiful ball-and-socket joint, which enables them to move freely in every direction. Now, these antennæ quite clearly serve several most important uses in ant life. They are the organs of speech in ants, as well as the organs of a special sense ; just as, with ourselves, the mouth is used equally for tasting and talking.

Darwin said with justice, indeed, that, considering its size, the brain of an ant was perhaps the
most marvellous piece of matter in the whole
universe; and its raw material of intelligence is
apparently supplied it most of all through the
mysterious antennæ.

No. 9 is a back view of the same head, with
the various jaws and mouthpieces expanded. It
shows very well the complicated nature of the
tongue, the palps, the shield, and so forth, and
also the powerful nipping jaws, with their closely
serrated and tooth-like edge — these last being
the weapons used in battle and in repelling the
attacks of large enemies. It also excellently exhibits the complex arrangement of the beautiful
jointed antennæ. The black spot in the centre
of the head above is the cut neck, or esophagus.
I advise you to look closely at the mouth-organs
in this microscopic drawing, and to compare them
with the corresponding parts in the wasp, illustrated by Mr. Enock in the last chapter.

Considering how important the antennæ are,
it will not surprise you to learn that the clean
little ants have a special instrument, like the bees
and wasps, for keeping these useful outgrowths
in proper order. The singular brush-and-comb
with which they clean them is shown in No. 10,
together with a smaller representation of the entire
leg on which it exists, so as to enable you to
see where the ant carries it. Ants, indeed, are
as fond of washing themselves as cats; and when
any accident happens to one, such as getting

smeared with honey, you will see the little
creature carefully getting rid of the foreign body
with her hairy legs, and paying particular attention to her precious antennæ. The mere existence of such developed brushes is sufficient to

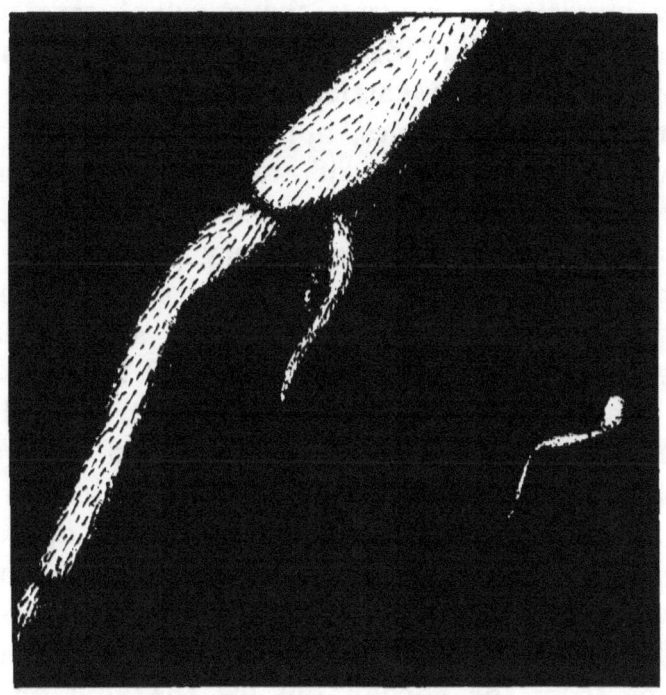

NO. 10.—THE ANT'S BRUSH-AND-COMB, FOR CLEANING
THE ANTENNÆ.

prove the immense importance of the organs they
clean to the bee-and-ant order.

The life-history of an ant falls into four periods
or ages: the egg, the grub, the pupa, and the
perfect insect. The eggs, which are very tiny,

are white or yellowish, and somewhat elongated; those observed by Sir John Lubbock, the great authority on ants, have taken a month or six weeks to hatch. The larvæ, like the young of bees and wasps, are white, legless grubs, narrow towards the head. The picture in No. 2, indeed, only imperfectly suggests the constant care with which they are tended by the nurses in early life; for they are carried about from room to room at different times, apparently to secure the exactly proper degree of warmth or moisture; and they are also often assorted in a sliding-scale of ages. "It is sometimes very curious to see them in my nests," says Sir John Lubbock, "arranged in groups according to size, so that they remind one of a school divided into five or six classes." After a longer or shorter period of grubhood, which differs in length in different species, they turn into pupæ, either in a cocoon or naked. It takes the insects three or four weeks, in the pupa form, to develop into full-grown ants; and even when they have finished, they are as helpless as babies, and could not escape from the cocoon but for the kind offices of the worker attendants. "It is pretty to see the older ants helping them to extricate themselves, carefully unfolding the legs and smoothing out the wings" of the males and females, "with truly feminine tenderness and delicacy." This utter helplessness of the young ant is very interesting for comparison with the case of man; for it is now known that nothing conduces to the final intellectual and moral supremacy of a race so much as the need for tending and carefully guarding the young ; the more complete the dependence of the offspring upon their elders, the finer and higher the ultimate development.

Ants are likewise great domesticators of various other animals ; indeed, as I have said before, they keep many more kinds of flocks and herds in confinement than we ourselves do. There is a funny little pallid creature, called Beckia, an active, bustling small thing, remotely resembling a minute earwig-larva, which runs in and out among the ants in great numbers, keeping its antennæ always in a state of perpetual vibration. The nests also harbour a queer, armour-plated white wood-louse, whose long Latin-German name I mercifully spare you ; and this strange beast toddles about quite familiarly among the ants in the galleries. Both kinds must have been developed in ants' nests from darker animals ; and both are blind, from long residence in the dark underground tunnels which they never quit ; their lightness of colour and the disappearance of their eyes tend alike to show that they and their ancestors have resided for countless ages in the homes of the ants. Yet no ant ever seems to take the slightest notice of them. Still, there they are, and the ants tolerate their presence ; while an unauthorised interloper, as Sir John Lubbock remarks, would at once be set upon and killed. The accomplished entomologist in question suggests that they may perhaps act as

scavengers, like the wild dogs of Constantinople
or the turkey-buzzard vultures of the West Indies
and South America. I have sometimes almost
been inclined to suspect, myself, that they may be
kept as totems, much as human savages domesticate one of their revered ancestral animals as an
object of worship.

In other cases the relation between the ants
and their domesticated animals is more distinctly
economical. For instance, there is a blind beetle
—most ant-cattle are blind from long residence
in the tunnels—which has actually lost the power
of feeding itself; but the ants feed it with their
own food, and then caress it with their antennæ,
apparently in order to make it give forth some
pleasant secretion. This secretion seems to be
poured out by a tuft of hairs at the base of the
beetle's hard wing-cases; these tufts of hair the
ants take into their mouths and lick all over with
the greatest relish. Some ant tribes even strike
up an alliance with other ants of a different
species, whose nest they frequent and whom they
follow in all their wanderings. Thus, there is a
very tiny yellow ant, known as Stenamma, which
takes up its abode in the galleries of the much
larger Horse Ants and Field Ants. When these
big friends change their quarters to a new nest,
as frequently happens, the tiny Stenammas accompany them, "running about among them," says
Sir John Lubbock, "and between their legs, tapping them inquisitively with their antennæ, and
even sometimes climbing on to their backs, as

if for a ride, while the large ants seem to take little notice of them. They almost seem to be the dogs, or perhaps the cats, of the ants." In yet another case, a wee parasitic kind makes its own small tunnels in and out among those of a much larger species, members of which cannot get at the petty robbers, because they are themselves too big to enter the minute galleries. The depredators are, therefore, quite safe, and make incursions into the nests of their bigger victims, whose larvæ they carry off and devour—"as if we had small dwarfs, about eighteen inches long, harbouring in the walls of our houses, and every now and then carrying off some of our children into their horrid dens."

When once one begins upon these fascinating insects, the difficulty is to know when to stop. But I have said enough, I hope, to suggest to you the extraordinary interest of the study of ant life. Even if observed in the most amateurish way, it affords one opportunities for endless amusing glimpses into the politics of a community full of comic episodes and tragic *dénouements*.

IX

A FROZEN WORLD

THE pond in the valley is a world by itself. So far as its inhabitants are concerned, indeed, it is the whole of the world. For a pond without an outlet is like an oceanic island; it is a system, a microcosm, a tiny society apart, shut off by impassable barriers from all else around it. As the sea severs Fiji or St. Helena from the great land-surface of the continents, so, and just as truly, the fields about this pond sever it from all other inhabited waters. The snails and roach and beetles that dwell in it know of no other world; to them, the pond is all; the shore that bounds it is the world's end; their own little patch of stagnant water is the universe.

A pond which empties itself into a river by means of a stream or brook is not quite so isolated. It has points of contact with the outer earth: it resembles rather a peninsula than an island: it is the analogue of Spain or Greece, not of Hawaii or Madeira. And you will see how important this distinction is if you remember that trout and stickleback and stone-loach and fresh-water mussels can ascend the river into the brook, and pass by the brook into the pond, which has thus a direct line of communication with all waters elsewhere, including even the great oceans. But the pond without an outlet cannot thus be peopled. Whatever inhabitants it possesses have come to it much more by pure chance. They are not able to walk overland from one pond to another ; they must be brought there somehow, by insignificant accidents. Regarded in this light, the original peopling of every pond in England is a problem in itself—a problem analogous in its own petty way to the problem of the peopling of oceanic islands.

That great and accomplished and ingenious naturalist, Mr. Alfred Russel Wallace, working in part upon lines long since laid down by Darwin, has shown us in detail how oceanic islands have in each case come to be peopled. He has shown us how they never contain any large indigenous land animals belonging to the great group of mammals—any deer or elephants or pigs or horses ; because mammals, being born alive, cannot, of course, be transported in the egg, and because the adult beasts could seldom be carried across great stretches of ocean by accident without perishing on the way of cold, hunger, or drowning. One can hardly imagine an antelope or a buffalo conveyed safely over sea by natural causes from Africa to the Cape Verdes, or from America to the Bermudas. As a matter of fact, therefore, the natural population

of oceanic islands (for I need hardly say I set
aside mere human agencies) consists almost entirely of birds blown across from the nearest continent, and their descendants; of reptiles, whose
small eggs can be transported in logs of wood
or broken trees by ocean currents ; of snails and
insects, whose still tinier spawn can be conveyed
for long distances by a thousand chances ; and of
such trees, herbs, or ferns as have very light seeds
or spores, easily whirled by storms (like thistle-
down), or else nuts or hard fruits which may be
wafted by sea-streams without damage to the
embryo. For the most part, also, the plants and
animals of oceanic islands resemble more or less
closely (with locally induced differences) those of
the nearest continent, or those of the land from
which the prevailing winds blow towards them,
or those of the country whence currents run most
direct to the particular island. They are waifs
and strays, stranded there by accident, and often
giving rise in process of time to special local
varieties or species.

Now, it is much the same with isolated ponds.
They acquire their first inhabitants by a series of
small accidents. Perhaps some water-bird from a
neighbouring lake or river alights on the sticky
mud of the bank, and brings casually on his
webbed feet a few clinging eggs of dace or chub,
a few fragments of the spawn of pond-snails or
water-beetles. Paddling about on the brink, he
rubs these off by mere chance on the mud, where
they hatch in time into the first colonists of the

new water-world. Perhaps, again, a heron drops a half-eaten fish into the water—a fish which is dead itself, but has adhering to its scales or gills a few small fresh-water crustaceans and mollusks. Perhaps a flood brings a minnow or two and a weed or two from a neighbouring stream; perhaps a wandering frog trails a seed on his feet from one pool to another. By a series of such accidents, each trivial in itself, an isolated pond acquires its inhabitants; and you will therefore often find two ponds close beside one another (but not connected by a stream), the plants and animals of which are nevertheless quite different.

Now, the pond in summer is one thing; the pond in winter is quite another. For just reflect what winter means to this little, isolated, self-contained community! The surface freezes over, and life in the mimic lake is all but suspended. Not an animal in it can rise to the top to breathe; not a particle of fresh oxygen can penetrate to the bottom. Under such circumstances, when you come to think of it, you might almost suppose life in the pond must cease altogether. But nature knows better. With her infinite cleverness, her infinite variety of resource, of adaptation to circumstances, she has invented a series of extraordinary devices for allowing all the plants and animals of a pond to retire in late autumn to its unfrozen depths, and there live a dormant existence till summer comes again. Taking them in the mass, we may say that the population sink down to the bottom in November or December, and surge up again in spring, though in most varied fashions.

Consider, once more, the curious set of circumstances which renders this singular plan feasible. Water freezes at 32 degrees Fahrenheit. For the most part, under normal conditions, the water at the top of the pond is the warmest, and that at the bottom coldest ; for the hot water, being expanded and lighter, rises to the surface, while the cold water, being contracted and heavier, sinks to the depths. If this relation remained unchanged throughout, when winter came, the coldest water would gradually congeal at the bottom of the pool : and so in time the whole pond would freeze solid. In that case, life in it would obviously be as impossible as in the ice of the frozen pole or in the glaciers of the Alps. But by a singular variation, just before water freezes, it begins to expand again, so that ice is lighter than water. Thus the ice as it forms rises to the surface, and leaves at the bottom a layer of slightly warmer water, some four or five degrees above freezing point. It is usual to point this fact out as a beautiful instance of special provision on the part of nature for the plants and animals which live in the ponds ; but to do so, I think, is to go just a step beyond our evidence. Nature does not fit all places alike for the development of life ; she does not fit the desert, for example, nor the interior of glaciers or frozen oceans, nor, for the matter of that, the rocks of the earth's mass ; nor does she try to fit living beings for such impossible situations. All we are really entitled to say is this—that the conditions for life *do* occur in ponds, owing to this habit of water, and that therefore special plants and animals have been adapted by nature to fulfil them.

The devices by which such plants and animals get over the difficulties of the situation, however, are sufficiently remarkable to satisfy the most exacting. Recollect that for some weeks together the entire pond may be frozen over, and that during that dreary time all animal or vegetable life at its surface must be inevitably destroyed. For hardly a plant or an animal can survive the actual freezing of its tissues. Nevertheless, as soon as winter sets in, the creatures which inhabit the pond feel the cold coming, and begin to govern themselves accordingly. A few, which are amphibious, migrate, it is true, to more comfortable quarters. Among these are the smaller newts or efts, which crawl ashore, and take refuge from the frost in crannies of rocks or walls, or in cool damp cellars. Most of the inhabitants of the pool, however, remain, and retire for warmth and safety to the depths. Even the amphibious frogs themselves, which have hopped ashore on their stout legs in spring, when they first emerged from their tadpole condition, now return for security to their native pond, bury themselves comfortably in the mud in the depths, and sleep in social clusters through the frozen season. They are not long enough and lithe enough to creep into crannies above ground like the newts; and with their soft smooth skins

O

and unprotected bodies they would almost inevitably be frozen to death if they remained in the
open.   On the bottom of the pond, however, they
huddle close and keep one another warm, so that
portions of the mud in the centre of the pool
consist almost of a living mass of frogs and other
drowsy animals.

Some of the larger pond-dwellers thus hibernate
in their own persons ; others, which are annuals,
so to speak, die off themselves at the approach
of winter, and leave only their eggs to vouch for
them and to continue the race on the return of
summer.   A few beetles and other insects split
the difference by hibernating in the pupa or chrysalis condition, when they would have to sleep
in any case, and emerging as full-fledged winged
forms at the end of the winter.   But on the whole
the commonest way is for the plant or animal
itself in its adult shape to lurk in the warm mud
of the bottom during the cold season.

In No. 1 we have an excellent illustration of this
most frequent type, in the person of the beautiful
pointed pond-snail, a common fresh-water mol-
lusk, with a shell so daintily pretty that if it did
not abound we would prize it for its delicate
transparent amber hue and its graceful tapering
form, resembling that of the loveliest exotics. This
pond-snail, though it lives in the water, is an air-
breather, and therefore it hangs habitually on the
surface of the pool, opening its lung-sac every now
and then to take in a fresh gulp of air, and looking
oddly upside down as it floats, shell downward,

in its normal position. It browses at times on the
submerged weeds in the pond ; but it has to come
to the surface at frequent intervals to breathe ;
though, in common with most aquatic air-breathers,
it can go a long time without a new store of oxygen, like a man when he dives, or a duck or swan
when it feeds on the bottom—of course to a much
greater degree, because the snail is cold-blooded ;
that is to say, in other words, needs much less
aëration. On a still evening in summer you will
often find the surface of the pond covered by
dozens of these pretty shells, each with its slimy
animal protruded, and each drinking in air at the
top by its open-mouthed lung-sac.

In winter, however, as you see in No. 2, our
pond-snail retires to the mud at the bottom, and
there quietly sleeps away the cold season. Being
a cold-blooded gentleman, he hibernates easily,
and his snug nest in the ooze, where he buries
himself two or three inches deep, leaves him relatively little exposed to the attacks of enemies.
Indeed, since the whole pond is then sleeping and
hibernating together, there is small risk of assault
till spring comes round again.

Now, it may sound odd at first hearing when I
tell you that what the animals thus do, the plants
do also. "What?" you will say. "A plant move
bodily from the surface of the water and bury
itself in the mud ! It seems almost incredible."
But the accompanying illustrations of one such
plant, the curled pond-weed, will show you that
the aquatic weeds take just as good care of

themselves against winter cold as the aquatic animals.

In No. 3 you see a shoot of curled pond-weed preparing to receive cold attacks at the approach of autumn. You may perhaps have noticed for

NO. I.—THE GREAT POND-SNAIL IN SUMMER.

yourself that almost all plants of stagnant waters tend to be freshest and most vigorous at the growing end—the upper portion ; while the lower and older part is usually more or less eaten away by browsing water beasties, or incrusted by parasites, or draggled and torn, or water-logged and mud-smeared. The really vital part of the plant at each moment is, as a rule, the top or growing shoot. Now, if the curled pond-weed were to let itself get overtaken bodily by winter, and its top branches or vigorous shoots frozen in the crust of ice which must soon coat the pond, it would be all

NO. 2.—THE GREAT POND-SNAIL IN WINTER.

up with it. To guard against this calamity, therefore, the plant has hit upon a dodge as clever in its way as that of our old friend the soldanella, which laid by fuel to melt the glacier ice in the Alpine springtide. Prevention, says the curled pond-weed, is better than cure. So, in No. 3, you catch it in the very act of getting ready certain

specialised detachable shoots, which are its liveliest
parts, and in which all the most active protoplasm
and chlorophyll (or living greenstuff of the plant)
are collected and laid by, much as food is laid by
in the bulb of a hyacinth or in the tuber of a
dahlia.   These shoots are, as it were, leafy bulbs,

NO. 3.—THE CURLED POND-WEED PRODUCING ITS WINTER SHOOTS.

meant to carry the life of the plant across the gulf
of winter.

In No. 4 we come upon the next act in this
curious and interesting vegetable drama.   Most
people regard plants as mere rooted things, with
no will of their own, and no power of movement.

In reality, plants, though usually more or less attached to the soil, have almost as many tricks and manners of their own as the vast mass of animals; they provide in the most ingenious and varied ways for the most diverse emergencies. The winter shoots of the curled pond-weed, for

NO. 4.—THE SHOOTS DETACHING THEMSELVES AND SINKING, BEFORE THE POND FREEZES.

example, carrying with them the hopes of the race for a future season, are deliberately arranged beforehand with a line of least resistance, a point of severance on the stem, at which in the fulness of time they peaceably detach themselves. You can note in the illustration how they have glided off

gently from the parent stalk, and are now sinking
by their own gravity to the warmer water of the
bottom, which practically never freezes in winter.
And the reason why they sink is that, being full
of rich living greenstuff, they are heavier than
the water, and heavier than the stem which pre-

NO. 5.—THE SHOOTS ROOTING AT THE BOTTOM WHILE THE POND
IS FROZEN.

viously floated them.   This stem has many air
cavities to keep it fairly erect and waving in the
water: but the winter shoots have none, so that
as soon as they detach themselves they sink of
their own mere weight to the bottom.   You may
notice that the leaves of deciduous trees in autumn

have similar lines, ordained beforehand, along
which they break off clean, so as not to tear or
injure the permanent tissues ; this is particularly
noticeable in the foliage of the horse-chestnut, and
also (in spring) in the common aralia, so often
grown as a drawing-room decoration.

NO. 6.—THE SHOOTS IN SPRING BEGINNING TO SPROUT AGAIN.

No. 5 continues the same series, and shows us
how the winter shoots, now sunk to the bottom,
bore a hole and root themselves in the soft mud
by their sharp, awl-like ends ; after which they
prepare to undergo their sleepy hibernation. They
are now essentially detached buds or cuttings,

analogous to those which the gardener artificially lops off and "strikes" in our gardens. Only, the gardener's cuttings have been rudely sliced off with a knife, after the crude human fashion, while those of the pond-weed have been neatly released without injury to the tissues, the separation being performed by an act of growth, with all the beautiful perfection that marks nature's handicraft.

In the soft slimy mud, the shoots of the curled pond-weed lie by during the frozen period, hearing the noise of the gliding skates above them, and suffering slightly at times from the chill of the water, but actually protected by the great-coat of ice from the severest effects of the hard weather. By-and-by, when spring comes again, however, the shoots begin to bud out, as you see in No. 6, and once more to produce the original type of pond-weed. The weed then continues to form leaves and stems, and finally to flower, which it does with a head or spike of queer little green blossoms, raised unobtrusively above the surface of the water. They are not pretty, because they do not depend upon animals for the transference of their pollen. I could tell you some curious things about these flowers, too, which find themselves far from insects, and destitute of attractive petals; so they have taken in despair to a quaint method of fertilisation by bombardment, so to speak—the stamens opening in calm weather, and dropping their pollen out on the saucer-like petals, whence the first high wind carries it off with a burst to the stigma or sensitive surface of the sister flowers. But that, though enticing, is another story, alien to the philosophy of the pond in winter. I will only add here that the pond-weed does not set its seeds very well, and that chances of dispersal are somewhat infrequent, so that irregular multiplication by these winter shoots has largely taken the place with it of normal multiplication by means of seedlings. At the same time, we must remember that no prudent plant can venture to depend for ever upon such apparent propagation by mere subdivision, which is not really (in any true sense) propagation at all, but is merely increased area of growth for the original parent, split up into many divergent personalities ; so that the curled pond-weed takes infinite pains all the same to flower when it can, and to discharge its pollen and disperse its seed as often as practicable. Only by seedlings, indeed (that is to say by fresh blood — truly new individuals), can the vigour of any stock be permanently secured.

Sometimes, again, the entire plant retires to the depths in winter, like the pond-snail. This is the case with that pretty floating aquatic lily, the water-soldier, whose lovely flowers make it a frequent favourite on ornamental waters. In summer it floats ; but when winter comes it sinks to the bottom, and there rests on the mud till spring returns again.

In No. 7 you see how another familiar and fascinating denizen of the pond, the little whirligig beetle, provides his winter quarters. The whirligig is one of the daintiest and most amusing of the inhabitants of our ponds. He is a small round

beetle, in shape like a grain of corn ; but as he is
intended to sport and circle on the surface of the
water in the broad sunshine, he is clad in glistening
mail of iridescent tints, gorgeous with bronze and
gold, to charm the eyes of his fastidious partner.
You seldom see whirligigs alone ; they generally
dart about in companies on the surface of some
calm little haven in the pond, a dozen at a time,
pirouetting in and out with most marvellous gyrations, yet never colliding or interfering with one
another.   I have often watched them for many
minutes together, wondering whether they would
not at last get in one another's way : but no ; at
each apparent meeting, they glide off in graceful
curves, and never touch or graze.   They go on
through figures more complicated than the Lancers
or Sir Roger de Coverley, now advancing, now
retreating, always in lines of sinuous beauty, without angularity or strain, and apparently without
premeditation ; yet never for a second do they
interfere with a neighbour's mazy dance, often as
they cross and recross each other's merry orbits.
Dear little playful things they seem, as if they
enjoyed existence like young lambs or children.
Sociable, alert, for ever gambolling, they treat life
as a saraband, but with a wonderfully keen eye for
approaching danger.   They look at times as if you
could catch them without trouble ; yet put down
your hand, and off they dart at once to the bottom,
or elude you by a quick and vigilant side movement, always on the curve, like a good skater or a
bicyclist.

This rapid skimming in curves or circles on the surface of the water is produced in a most interesting way by the co-operation of the various pairs of legs, which I can best explain by the analogy of the bicycle. The two shorter and active hind-legs produce the quick forward dart, just as the main

NO. 7.—THE WHIRLIGIG BEETLE IN SUMMER, DANCING.

motion of the cycle is given it by the back wheel; the longer front legs act like the front wheel of the cycle in altering the direction; one of them is jerked out to right or left, rudderwise, and gives the desired amount of curve to the resulting motion according to the will and necessities of the insect.

The steering of a Canadian canoe comes very near
it. Anybody who has sculled or rowed, indeed,
knows well the extraordinary ease with which
a boat can be shored off instantaneously from
another, or the marvellous way in which gliding
curves can be produced on the almost unresisting
surface of the water. The whirligig beetle has a
perfect steering apparatus in his long and ex-
tensible fore-legs, and by their means he performs unceasingly his play of merry and intricate
evolutions.

When whirligigs are alarmed, however, they dive
below the surface as one of a pair is doing in
No. 7, and carry down with them a large bubble of
air, for breathing purposes, entangled in the joints
of their complicated legs and the under parts of
their bodies. On this quaint sublacustrine balloon
they subsist for breathing till the danger is past
and they can come to the top again.

Early in April, when the weather is fine, you
begin to see the whirligig beetles dancing in and
out in companies, like so many water-fairies, on
the still top of the pond. They prefer calm water;
when the wind drives little ripples to the eastern
end of the pool, you will find them practising their
aquatic gymnastics under lee of the shore on the
western side; when an east wind ruffles the western
border, you will find them gyrating and interlacing,
coquetting and pirouetting, by the calmer eastern
shallows. As they move in their whirls, they form
little transient circles on the water's top, which
spread concentrically; and the mutual interference

of these widening waves is almost as interesting at times as the astonishing velocity and certainty of movement in the beetles themselves. So, all summer long, they continue their wild career, seeming to earn their livelihood easily by amusing them-

NO. 8.—WHIRLIGIG BEETLES IN WINTER, SLEEPING.

selves. But as soon as winter approaches, a change comes o'er the spirit of their dream. They retire to the depths, as you may observe in No. 8, and bury themselves in the mud while the pond is frozen over. During this period they indulge in a good long nap of some five or six months, and, awaking refreshed in April, come to the surface once more, where they begin their gyratory antics all over again, *da capo*. It is a merry life ; and though the whirligig can fly, which he does occasionally, 'tis no wonder he prefers his skimming existence on the still glassy sheet of his native waters.

The two larger British water-beetles, which are such favourite objects in the aquariums of young naturalists, do not lead quite so exclusively aquatic a life ; they pass their youth as larvæ in the pond, and they return to it in their full-winged or beetle stage, being most expert divers ; but they both retire to dry land to undergo their metamorphosis into a chrysalis, and they spend their time in the pupa-case in a hollow in the ground. Something similar occurs with many other aquatic animals, which are thus conjectured to be the descendants of terrestrial ancestors, whom the struggle for life has forced to embrace the easier opening afforded by the waters.

In this respect, that rather rare and beautiful little water-plant, the frogbit, shown in No. 9, has a life-history not unlike the career of the water-beetles. It is a quaint and pretty herb, which never roots itself in the mud, like the curled pond-weed, but floats freely about on the surface, allowing its long roots to hang down like streamers into the water beneath it. The short stem or stock is submerged ; the leaves expand themselves freely and loll on the surface. Like most other floating water-leaves which thus support themselves on the top of the water, they are almost circular in form

—a type familiar to all of us in the white and yellow water-lily, and also in the beautiful little fringed limnanthemum. The reason why floating

NO. 9.—THE FROGBIT IN SUMMER, FLOWERING.

leaves assume this circular shape is easy to perceive ; they need no stout stalk to support them; like aerial foliage, the water serving to float them on its surface ; and as they find the whole surrounding

P

space free from competition, with no other plants
to interfere with them, as in the crowded meadows
and hedgerows of the land, they spread freely in
the sunshine on every side, drinking in from the
air the carbonic acid which is the chief food of
plants. In short, the round shape is that which
foliage naturally assumes when there is no competition, no architectural or engineering difficulty,
plenty of food and plenty of sunshine.

The frogbit as a whole, then, is not submerged
like the curled pond-weed; it floats, not rooted,
but free. Yet when it comes to flowering, it has
to quit the water, just like the great water-beetles,
and emerge upon the open air above, so as to
expose its flowers to the fertilising insects. These
flowers are extremely delicate and beautiful, with
three papery white petals, and a yellow centre;
they make the plant a real ornament to all the
ponds where it fixes its residence. The males and
females grow on separate plants, and aquatic flies
act as their ambassadors. Such is the summer life
of the frogbit, while fair weather lasts; but, like
all other pond denizens, it has to reckon in the
end with the frozen season.

It does so in a way slightly different from, though
analogous to, that of the curled pond-weed. No. 10
shows you the frogbit after the flowering season is
over, when it begins to anticipate the approach of
winter. It then sends out slender runners, like
those of the strawberry vine, on the end of each of
which is formed a winter bud, which answers to
the winter shoots of the curled pond-weed. By-

and-by the pond will freeze, and the floating leaves of the frogbit will be frozen and killed with it. But the prudent plant provides for its own survival

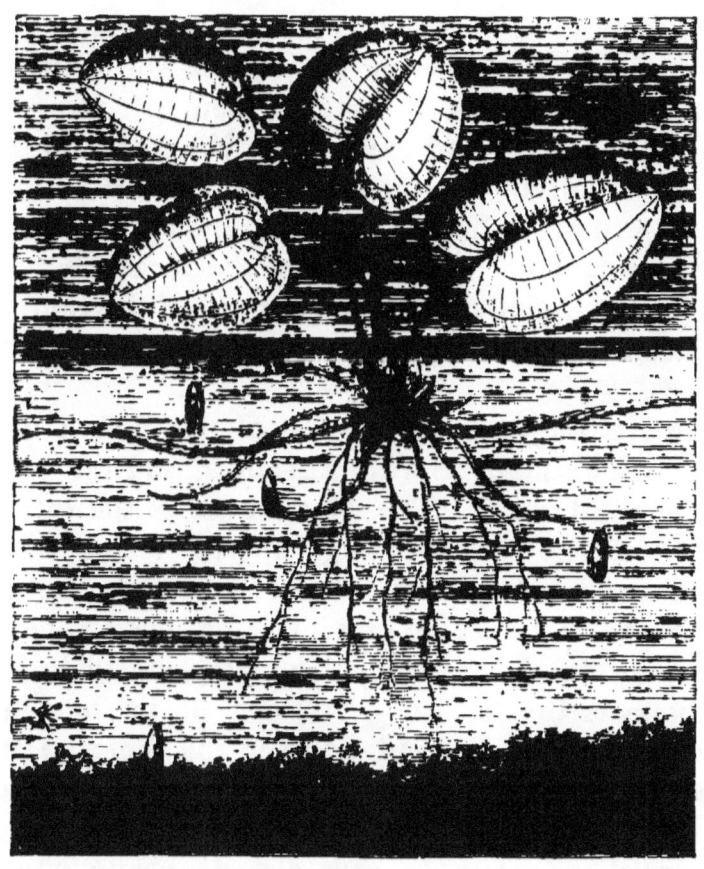

NO. 10.—THE FROGBIT DETACHING ITS WINTER BUDS, WHICH SINK TO THE BOTTOM.

in the person of its offshoots, which are not its young, but integral parts of its own individuality. It fills them with starch and other rich foodstuffs for growth next season. About the time when the pond grows cool, the buds detach themselves, like the winter shoots of the pond-weed, and slowly descend by their own weight to the bottom. But they do not root themselves there, as the pond-weed shoots did; they merely lie by, like the whirligig beetles, as you can see one of them preparing to do in the left-hand corner of No. 10. All the living material is drained from the leaves into these winter bulbs. The pond freezes over, and the remnant of the floating leaves decay; but the bulbs lurk quietly in the warm mud of the bottom, protected by a covering of close-fitting scale-leaves.

In No. 11 we learn the end of this quaint little domestic drama. Spring has come, and the pond has thawed again. The winter buds of the frogbit now undergo certain spongy internal changes, due to warmth and growth, which make them lighter —lessen their specific gravity. Air-cells are developed in them. So they begin to rise again like bubbles to the surface. You can see in the illustration one bud still entangled in the slime on the bottom; another just starting to emerge; a third rising; and a fourth and fifth on the surface of the pool. Two more have already risen; one of these is just putting forth its first few kidney-shaped leaves; another has now grown pretty strong, and is sending out a runner, from which a third little plant is even beginning to develop. In time, hundreds of such runners are sent forth in every direction, till the surface of the pond, in suitable places, is covered with a network of tangled and interlacing frogbits. They always seem to me in this way the plant-counterparts of the whirligig beetles ; and it is because of this queer analogy in their mode of life that I have figured the two here in such close connection.

NO. II.—THE BUDS RISING AGAIN IN SPRING, AND SPROUTING INTO A NETWORK.

Indeed, I hope I have now begun to make it clear to you that the difference of habit between plants and animals is not nearly so vast as most people imagine. It is usual to think of plants as merely passively existing. I have tried, here and elsewhere, to lay stress rather upon the moments in life when plants are *doing something*, and thus to suggest to my readers the close resemblance which really exists between their activities and those of animals. The more you watch plants, the more will you find how much this is true. And in a case like that of a pond frozen in winter, where both groups have to meet and face the self-same difficulty, it is odd to note how exactly similar are the various devices by which either group has succeeded in surmounting it.

When you skate carelessly over the frozen pond in winter, you never perhaps reflect upon all the wealth of varied life that lies asleep beneath your feet. But it is there in abundance. The smaller newt, to be sure, has gone ashore to hibernate : but his great crested brother lurks somnolent in the mud, like a torpid bear or a sleeping dormouse. Frogs huddle buried in close packed groups at the centre, massed together in the soft ooze for warmth and company. Many kinds of aquatic snails slumber peaceably hard by, with various beetles beside the whirligigs. As for eggs and spawn and larvæ or pupæ, as well as petty crustaceans, you could count them by the dozen. Seeds are there, too, and buried plants of water-crowfoot, and winter shoots and winter buds, and a whole world of skulkers. The pond seems dead, if you look only at its hard and frozen top ; but in its depths it encloses for kind after kind the manifold hope of a glorious resurrection. Let May but come back with a few genial suns, and forthwith, the water-crowfoot spreads its white sheet of tender bloom ; the whirligig dances anew ; the newts acquire their red and orange spots and their decorative crests ; strange long-legged creatures stalk on stilts over the glass of the calm bays, and tadpoles swarm black and fat in the basking shallows. The pond, it seems, was not dead but sleeping. Spring sounds its clarion note, and all nature is alive again.

X

BRITISH BLOODSUCKERS

I WRITE this title with peculiar pleasure, because it is so nice to be able for once to apply it literally. With its figurative use I am already too familiar. In some tropical countries the free-born Britons who are sent out in the Government employment to protect the natives or the coolies or the negroes, as the case may be, from their aggressive brethren, are commonly known to their planter neighbours as "British bloodsuckers" —apparently because, like most other members of Civil Services elsewhere (except the Turkish), they get paid for their services. This use of the phrase is so well known to me, even as applied to myself, that I rejoice in being able to employ it here, without political prejudice of any sort, with reference to the habits of the mosquito and the horse-fly. Nobody, I suppose, is interested to deny that mosquitoes and horse-flies *do* suck blood ; nobody feels the faintest sympathy for the misdeeds of those sanguinary and unpleasant creatures. Now, it is always delightful to find a lawful outlet for our evil passions : all the world turns out to hunt a mad dog. I love to flick the heads off tall thistles

with my stick as I pass, and salve my scruples with
the thought that they are the deadly enemies of
the agricultural interest. If there were no thistles,
there would be nothing in the shape of a large
and conspicuous flower whose head one could
knock off with a clear conscience.

But at the very outset, I foresee a destructive criticism. " The mosquito," you will say, " is
not a *British* bloodsucker." Pardon me ; there,
you labour under a misapprehension. Everybody
knows that there are gnats in England. Well, a
gnat is a mosquito and a mosquito is a gnat. Like
our old friend, Colonel Clay, they are the same
gentleman under two different aliases. Or, rather,
since it is only the female insect that bites, and
only the bite that much concerns humanity, I ought
perhaps to say the same lady. The difference of
name is a mere question of nomenclature, and also
(as with many other aliases) a question of where
we happen to meet them. When a mosquito is
seen in England, he or she is called a gnat ; when
a gnat is seen in Italy or Egypt, he or she is
called a mosquito. But, as this is a fundamental
point to our subject, I think we had better clear it
up once for all before we go any farther. It is
not much use talking about mosquitoes unless we
really decide what particular creature it is that we
are talking about.

There is not one kind of gnat, or one kind of
mosquito, but several kinds of them ; and both
names are loosely applied in conversation to cover
a large variety of related small flies, almost all of

them members of the genus Culex. The one point
of similarity between the whole lot lies in the fact
that they all suck blood ; whenever a blood-sucking
culex is lighted upon in England it is called a
gnat ; while whenever one is found in any other
part of Europe, Asia, Africa, or America, we say
it is a mosquito. That is just a piece of the well-
known British arrogance ; they will not admit that
there are such venomous beasts as mosquitoes in
England, and therefore, when found, they call
them by another name, and fancy they have got
rid of them. As a matter of fact, mosquitoes of
one sort or another occur in most countries, if not
in all the world ; they are most numerous, it is
true, in the tropics and in warm districts generally ; but they also abound in Canada, Siberia,
Russia, and Lapland. Even in the Arctic regions,
they come out in swarms during the short summer ;
and wherever ponds or stagnant waters abound in
Finland or Alaska, they bite quite as successfully
and industriously while they last as in Ceylon or
Jamaica. At least a hundred and fifty kinds are
" known to science," and of these, no fewer than
thirty-five occur in Europe. There are nine in
Britain. Most of the European species bite quite
hard enough to be popularly ranked as mosquitoes ;
the remainder are called by the general and indefinite name of flies—a vague term which covers
as large an acreage of evil as charity.

In hot summers, you will often read in the
papers a loud complaint that " mosquitoes have
made their appearance in England," most often in

the neighbourhood of the London docks ; and this
supposed importation of venomous foreign insects
is usually set down to the arrival of some steamer
from Bombay or New Orleans. The papers might
almost as well chronicle the "arrival" of the cock-
roach or of the common house-fly. There are
always mosquitoes in England ; and they bite worse
in very hot weather. Occasionally, no doubt,
some stray Mediterranean or American gnat, rather
hungrier than usual, does cross over in water in
the larval form and effect a lodgment in London
for a week or two ; but only a skilled entomologist
could distinguish him from a native, after careful
examination. Let it be granted then, as Euclid
says, that there is no essential difference between
a gnat and a mosquito, and let us admit that the
same name is applied in both cases to a large
variety of distinct but closely related species.
After which preliminary clearing of the ground,
we will proceed quietly to the detailed description
of one such typical bloodsucker.

In justice to India, however, I ought perhaps to
add that the particular mosquito chosen for illustration by Mr. Enock is not itself a native Briton,
but an inhabitant of India. It is thus only British
in the wider sense of being a denizen of her
Majesty's dominions, on which the sun never sets,
and the buzz of the mosquito never ceases. On
the other hand, it differs so slightly from the
commonest English gnat that nobody but a trained
entomologist could ever detect the difference ; and
even he could only discover it in the adult insect

by minute variations in the antennæ and other almost microscopic peculiarities.   Indeed, if I hadn't told you this was an Indian mosquito, you would never have discovered that it wasn't a Fenland gnat.

The mosquito is in a certain sense an amphibious animal ; that is to say, during the course of its life, it has tried both land and water.   It begins existence as an aquatic creature, and only steps

NO. 1.—MOSQUITO'S EGG-RAFT, SEEN SIDEWAYS.

ashore at last to fly in the open air when it has arrived at its adult form and days of discretion. The mother mosquito, flitting in a cloud-like swarm of her kind, haunts for the most part moist and watery spots in thick woods or marshes, and lays her tiny eggs on the surface of some pool or stagnant water.   They are deposited one by one, and then glued together with a glutinous secretion into a little raft or boat, shown in No. 1, which floats about freely on the pond or puddle.   It looks just like the conventional representations of the "ark of bulrushes" provided for the infant Moses. An industrious mother will lay some two or three hundred such eggs in a season, so that we need not wonder at the great columns of mosquitoes that often appear in damp places in summer. No. 2 shows the same raft seen from above, and ex-

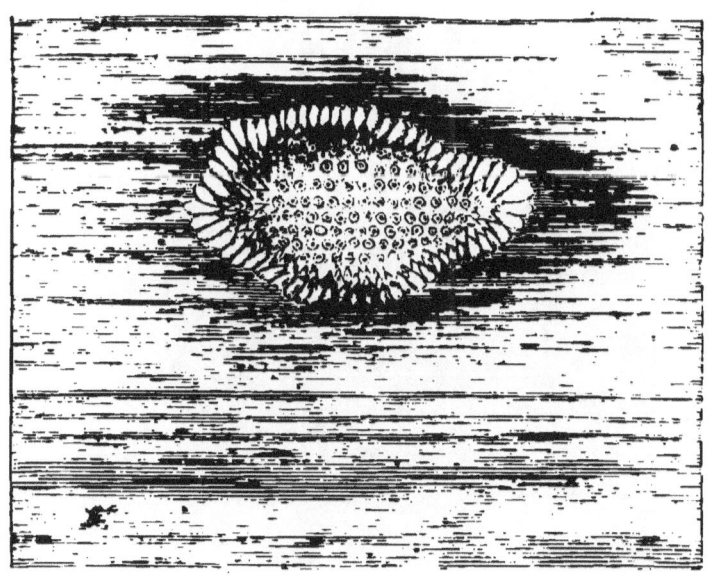

NO. 2.—THE MOSQUITO'S EGG-RAFT, SEEN FROM ABOVE.

cellently illustrates its admirable boat-shaped or saucer-shaped construction.

After about three days' time, the eggs begin to hatch, and the active little larvæ escape, wriggling, into the water. No. 3, which is enlarged forty diameters, exhibits the stages of the hatching process. A sort of lid or door at the lower end of the floating egg opens downward into the water, and the young mosquito slides off with a jerk of the tail into its native marshes. Almost everybody who has travelled in Asia, Africa, or America, must be familiar with these little brown darting larvæ,

NO. 3.—THE EGGS HATCHING, AND YOUNG MOSQUITOES ESCAPING.

which occur abundantly in the soft water in jugs and wash-hand basins. Brown, I say roughly, because they look so at a casual glance ; but if you examine them more closely you will see that they are rather delicately green, and often mottled. It is not easy to catch them, however, so quickly do they wriggle ; you try to put your hand on them, and they slip through your fingers ; you have caught one now, and, hi presto ! before you know it, he is twirling off to the other side and disporting himself gaily in aquatic gambols.

Nevertheless, he is a creature well worth observing, this larva. Get him still under the microscope (which is no easy matter— to insure it, you must supply him with only the tiniest possible drop of water) and you will then perceive that he has a distinct head, with two large dark eyes,

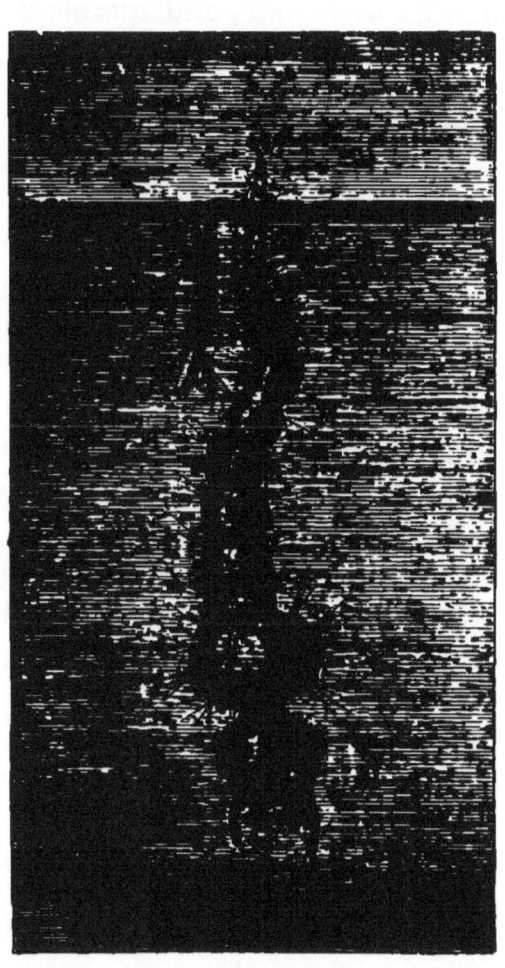

NO. 4—THE MOSQUITO-LARVA IN HIS FAVOURITE ACT OF STANDING ON HIS HEAD AND BREATHING.

and that behind it comes a globular body, and then a tail of several quickly-moving segments. No. 4

is a portrait of the larva in his full-grown stage,
near the surface of the water. He is about
half an inch long, and nimble as a squirrel. You
will observe on his head a sort of big moustache,
set with several smaller bristles. This moustache
(which consists for science of a pair of mandibles)
is kept always in constant and rapid motion ; its
use is to create an eddy or continuous current
of water; which brings very tiny animals and
other objects of food within reach of the voracious
larva's mouth ; for young or old, your mosquito
is invariably a hungry subject. In point of fact,
you may say that these hairy organs are the
equivalents of hands with which the larva feeds
himself. They vibrate ceaselessly.

At the opposite end of the body, you will
observe, there are two other organs, both equally
interesting. One of them, which goes straight
up to the surface of the water, and protrudes
above it, is the larva's breathing-tube ; for the
mosquito breathes, at this stage, not with his head
but with his tail; this ingenious mechanism I
will explain further presently. The other organ,
which in the illustration (No. 4) goes off to the
left, and has four loose ends visible, serves its
owner as a fin and rudder. It is the chief organ
of locomotion—the oar or screw by whose means
the larva darts with lightning speed through the
water, and alters his direction with such startling
rapidity. You will note that it is not unlike the
screw of a steamer, and it answers for the animal
the same general purpose. How effectual it is

as a locomotive device everybody knows who has once tried chivvying a few healthy mosquito larvæ round the brimming sea of his bedroom basin.

The breathing - tube deserves a little longer notice. By its means air is conveyed direct into the internal air-channels of the insect, which do not form lungs, but ramify like arteries all over the body. We carry our blood to the lungs to be aerated; the insects carry the oxygen to the blood. To take in air, the larva frequently rises to near the surface, as you see him doing in No. 4; then he stands on his head, cocks up his tail, and pushes out his air-tube.

NO. 5.—THE LARVA'S BREATHING-TUBF, CLOSED AND OPEN.

Indeed, when at rest this is his usual attitude. No. 5, which, of course, is very highly magnified, shows his tail in the act of taking in a good gulp of oxygen. The little valves, or doors, which cover the air-tube are here opened radially, and the larva is breathing. To the right you see the position of the tube after he has taken in a long draught of air (just like a whale or a porpoise) and is darting to the depths again. The tiny valves or doors are now closed, so that no water can get in; the larva will go on upon the air thus stored till all of it is exhausted; he will then rise once more to the surface, let out the breath loaded with carbonic acid, and draw in a fresh stock again for future use.

NO. 6.—THE PUPA OR CHRYSALIS, BREATHING THROUGH
TWO HORN-LIKE TUBES.

The young mosquito remains in the larval form for about a fortnight or three weeks, during the course of which time he moults thrice. As soon as he is full-grown, he becomes a pupa or chrysalis—lies by, so to speak, while he is changing into the winged condition. No. 6 is a faithful portrait of the mosquito in this age of transition. (I borrow the last phrase from the journalists.)

Within the pupa-case, which is smaller than the larva, the insect is bent double; in this apparently uncomfortable position, it begins to develop the wings, the legs, and the blood-sucking apparatus of the perfect mosquito. Nevertheless, ill-adapted as such a shape might seem for locomotion—with one's head tucked under, and one's eyes looking downward—the mosquito in the pupa continues to move about freely, instead of taking life meanwhile in the spirit of a mummy in the mummy-case. By way of change, however, he now eats nothing—having, in fact, no mouth to eat with. But the most wonderful thing of all is the alteration in his method of breathing. The pupa no longer breathes with its tail, but with the front part of its body, where two little horn-shaped tubes are developed for the purpose. You can see them in the illustration (No. 6), which is taken at the moment when the active and loco-motive pupa has just come to the surface to breathe, and is floating, back up, and head doubled under downward, in a most constrained position. The attitude reminds one of nothing so much as that of a bull, with his head between his legs, rushing forward to attack one. You can see through the pupa-case the great dark eyes and the rudiments of the legs as they form below it.

No. 7 exhibits very prettily the next stage in this short eventful history—the emergence of a female mosquito from her dressing-gown or pupa-case. She looks like a lady coming out of her ball-dress. As the pupa grows older, the skin or case stands off of itself from the animal within, by a sort of strange internal shrinkage, and a layer of air is thus formed between case and

NO. 7.—THE FEMALE MOSQUITO ABANDONING HER PUPA CASE.

occupant. This causes the whole apparatus to float to the surface, and enables the winged fly to make an effective exit. The new mosquito, looking still very hump-backed, and distinctly crouching, breaks through the top of the pupa-case (which opens by a slit), raises herself feebly and awkwardly on her spindle shanks, and withdraws her tail from its swathing bandage. She has grown meanwhile into a very different creature from the aquatic larva : observe her long plumed antennæ, her curious mouth-organs, her six hairy legs, and her delicate gauze-like wings, all of them wholly distinct from her former self, and utterly unrepresented by anything in the swimming insect. It is a marvellous transformation this, from a darting aquatic with rudder and tail, to a flying terrestrial and aerial animal, with legs and wings and manifold adapted appendages. At first, one would say, the new-fledged mosquito can hardly know herself.

In nature, however, nothing is ever wasted. The pupa-case, you would suppose, is now quite useless. Not a bit of it. Our lady utilises it at once as a boat to float upon. She plants her long legs upon it gingerly, as you see in No. 8, where you can still make out the shape of the tail and the horn-like breathing-tubes of the pupa. Thus does she rise on stepping-stones of her dead self to higher things, in a more literal sense than the poet contemplated. You observe her above, in her natural size, and below much magnified. Notice her beautiful gauzy wings, marked with hairy veins, her pretty plume-like antennæ, her spider-like jointed legs, and her hump of a body. She stands now, irresolute, meditating flight and wondering whether she dare unfold her light pinions to the breeze. Soon, confidence and strength will come to her ;

she will plim them on the summer air, and float away carelessly, seeking whom she may devour.

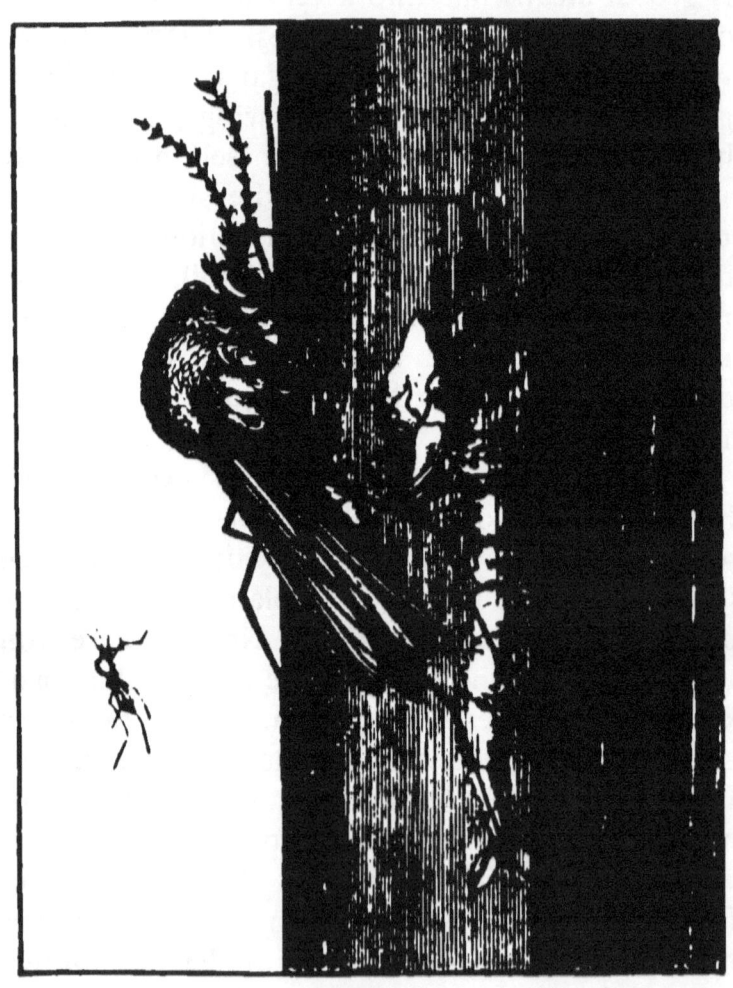

NO. 8.—THE FEMALE MOSQUITO MAKING A BOAT OF HER CAST-OFF SKIN.

All this is what happens to a successful insect. But often the boat fails; the young wings get wetted; the mosquito cannot spread them; and so she is drowned in the very element which till now was the only place where she could support existence.

And here I must say a word in favour of the male as against the female mosquito. In most species, and certainly in the commonest British gnat, the male fly never sucks blood at all, but passes an idyllic vegetarian existence, which might excite the warmest praise from Mr. Bernard Shaw, in sipping the harmless nectar of flowers. He has, in point of fact, no weapon to attack us with. He is an unarmed honey-sucker.

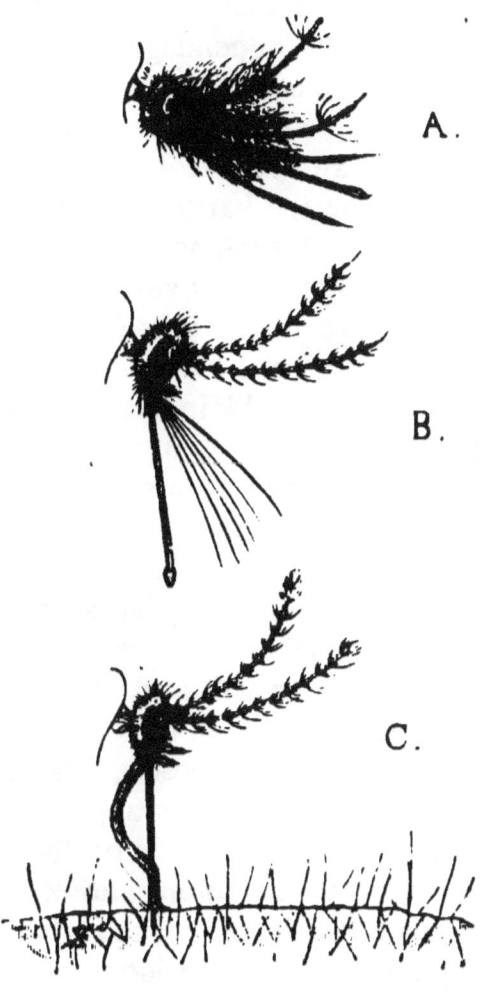

NO. 9.—HEADS OF MOSQUITOES; A, THE WHISKERED MALE; B, THE BLOOD-SUCKING FEMALE, WITH LANCETS EXPOSED; C, THE FEMALE, BITING A HUMAN HAND.

But the female is very differently minded—a Mes-

salina or a Brinvilliers, incongruously wedded to a
vegetarian innocent. Even the very forms of the
head and its appendages are quite different in the
two sexes in adaptation to these marked differences
of habit. No. 9 shows us the varieties of form in
the male and female at a glance. Above (in Fig.
A) we have the harmless vegetarian male. Observe
his innocent sucking mouth, his bushy beard, his
lack of sting, his obvious air of general respecta-
bility. He might pass for a pure and blameless
ratepayer. But I must be more definitely scientific,
perhaps, and add in clearer language that what I
call his beard is really the antennæ. These consist of fourteen joints each, fitted with delicate
circlets of hair ; and the hairs in the male are so
long and tufted as to give him in this matter a
feathery and military appearance, wholly alien to
his real mildness of nature. Look close at his
head and you will find it is provided with three
sets of organs—first, the tufted antennæ ; second, a
single sucking proboscis, adapted for quiet flower-
hunting and nectar-eating ; third, a pair of long
palps, one on each side of the proboscis.

Now, beneath him, marked B, we get the head
of his faithful spouse, the abandoned blood-sucking mosquito, which looks at first sight, I confess,
much more simple and harmless. Its antennæ
have shorter and less bristling hairs ; its proboscis
seems quiet enough ; and its palps are reduced to
two mere horns or knobs, not a quarter the length
of the bristly husband's, on each side of the proboscis. But notice in front of all that she has

five long lancets, guarded by an upper lip, which
do not answer to anything at all in her husband's
economy. Those five lancets, with their serrated
points, are the awls or piercers with which she
penetrates the skin of men or cattle. They cor-
respond to the mandibles, maxillæ, and tongue,
which I shall explain hereafter in the mouth of
the gadfly. How they work you can observe in
the lowest figure, C. Here you have a bit of the
hand of a human subject—not to put too fine
a point upon it (which is the besetting sin of
mosquitoes), the artist's. He has delivered himself up to be experimented on in the interests
of science. The sharp lancets have been driven
through the skin into the soft tissue beneath, and
the bent proboscis is now engaged in sucking up
the blood that oozes from it. If that were all, it
would be bad enough ; but not content with that,
the mosquito, for some mysterious reason, also
injects a drop of some irritant fluid. I have never
been able to see that this proceeding does her any
good, but it is irritating to us ; and that, perhaps,
is quite sufficient for the ill-tempered mosquito.

Owing to the habits of the larva, mosquitoes are
of course exceptionally abundant in marshy places.
They were formerly common in the Fen district of
England, but the draining of the fens has now
almost got rid of them, as it has also of the fever-
and-ague microbe.

As a rule, mosquitoes are nocturnal animals,
though in dark woods, and also in very swampy
districts, they often bite quite as badly through the

daytime as at night.   But when evening falls, and
all else is still, then wander forth these sons (or
daughters) of Belial, flown with insolence and
blood.   "What time the grey fly winds her sultry
horn," says Milton ; and that sultry horn is almost
more annoying than the bite which it precedes.
You lie coiled within your mosquito - curtains,
wooing sweet sleep with appropriate reflections,
when suddenly, by your ear, comes that still small
voice, so vastly more pungent and more irritating
than the voice of conscience.   You light a candle,
and proceed to hunt for the unwelcome intruder.
As if by magic, as you strike your match, that
mosquito disappears, and you look in vain through
every fold and cranny of the thin gauze curtains.
At last you give it up, and lie down again, when
straightway, "z-z-z-z," the humming at your ear
commences once more, and you begin the unequal
contest all over again.   It is a war of extermina-
tion on either side—you thirst for her life, and
she thirsts for your blood.   No peace is possible
till one or other combatant is finally satisfied.

You can best observe the mosquito in action,
however, by letting one settle undisturbed on the
back of your hand, and waiting while she fills
herself with your blood ; you can easily watch her
doing so with a pocket lens.   Like the old lady
in "Pickwick," she is soon "swelling wisibly."   She
gorges herself with blood, indeed, which she straightway digests, assimilates, and converts into the 300
eggs aforesaid.   But if, while she is sucking, you
gently and unobtrusively tighten the skin of your

hand by clenching your fist hard, you will find that she cannot any longer withdraw her mandibles; they are caught fast in your flesh by their own harpoon-like teeth, and there she must stop accordingly till you choose to release her. If you then kill her in the usual manner, by a smart slap of the hand, you will see that she is literally full of blood, having sucked a good drop of it.

The humming sound itself by which the mosquito announces her approaching visit is produced in two distinct manners. The deeper notes which go to make up her droning song are due to the rapid vibration of the female insect's wings as she flies; and these vibrations are found by means of a siren (an instrument which measures the frequency of the waves in notes) to amount to about 3000 in a minute. The mosquito's wings must, therefore, move with this extraordinary rapidity, which sufficiently accounts for the difficulty we have in catching one. But the higher and shriller notes of the complex melody are due to special stridulating organs situated like little drums on the openings of the air-tubes; for the adult mosquito breathes no longer by one or two air-entrances on the tail or back, like the larva, but by a number of spiracles, as they are called, arranged in rows along the sides of the body, and communicating with the network of internal air-chambers. The curious mosquito music thus generated by the little drums serves almost beyond a doubt as a means of attracting male mosquitoes, for it is known that the long hairs on the antennæ of the males, shown in No. 9,

Fig. A, vibrate sympathetically in unison with the notes of a tuning-fork, within the range of the sounds emitted by the female. In other words, hairs and drums just answer to one another. We may, therefore, reasonably conclude that the female sings in order to please and attract her wandering mate, and that the antennæ of the male are organs of hearing which catch and respond to the buzzing music she pours forth for her lover's ears. A whole swarm of gnats can be brought down, indeed, by uttering the appropriate note of the race ; you can call them somewhat as you can call male glow-worms by showing a light which they mistake for the female.

NO. 10.—THE GADFLY, NATURAL SIZE.

A much larger and more powerful British bloodsucker than the mosquito, again, is the gadfly or horse-fly, whose life-size portrait Mr. Enock has drawn for us in No. 10. Most people know this fearsome beast well in the fields in summer. He has a trick of settling on the back of one's neck, and making a hole in one's skin with his sharp mandibles ; after which he quietly sucks one's blood almost without one's perceiving him. Horses in pastures are often terribly troubled by these persistent creatures, which make no noise, but creep

silently up and settle on the most exposed parts of
the legs and flanks.   They are very voracious, and
manage to devour an amount of blood which is
truly surprising.

A little examination of the gadfly will show you,
too, one important point in which it and all other
true flies differ from the bees, wasps, butterflies,
and the vast mass of ordinary insects.   All the
other races have four wings, and I showed you in
the case of the wasp the beautiful mechanism of
hooks and grooves by which the fore and hind
wings are often locked together in one great group,
so as to insure uniformity and fixity in flying.
Among the true flies, however, including not only
the house-fly and the meat-fly, but also the gadflies
and the mosquitoes, only one pair of wings—the
front pair—is ever developed.   The second or hind
pair is feebly represented by a couple of tiny rudimentary wings, known as poisers or balancers,
which you can just make out in the sketch, like a
couple of stalked knobs, in the space between the
true wings and the tail or abdomen.   It is pretty
clear that the common ancestor of all these two-
winged flies must have had four wings, like the
rest of the great class to which he belonged ; but
he found it in some way more convenient for his
purpose to get rid of one pair, and he has handed
down that singular modification of structure to all
his descendants.   Yet whenever an organ or set
of organs is suppressed in this way, it almost always
happens that rudiments or relics of the suppressed
part remain to the latest generations ; and thus the

true flies still retain, in most cases, the two tiny poisers or balancers, just to remind us of their descent from four-winged ancestors. Nature has no habit more interesting than this retention of parts long since disused or almost disused ; by their aid we are able to trace the genealogy of plants and animals.

In No. 11 we have a dissected view of the mouth-organs and blood-sucking apparatus of the gadfly, immensely enlarged, so as to show in detail the minute structure. In life, all these separate parts are combined together into a compound sucker (commonly called the proboscis), which forms practically a single tube or sheath ; they are dissected out here for facility of comprehension. The longest part, marked LA in the sketch, is the *labium* or lower lip, which makes up the mass of the tube ; it ends in two soft finger-like pads, which are fleshy in texture, and which enable it to fix itself firmly

NO. 11.—THE GADFLY'S LANCETS, WITH OTHER PARTS OF THE PROBOSCIS.

(like a camel's foot) on the skin of the victim. The grooved and dagger-shaped organ, marked LBR, is the *labrum*, or upper lip ; and the tube or sheath formed by the shutting together of these two parts encloses all the other organs. Combined, they form a trunk or proboscis, not unlike that of the elephant. But the elephant is not a bloodsucker ; his trunk encircles no dangerous cutting weapon. It is otherwise with the gadfly, which has a pair of sharp knives within, for lancing the thick skin of its unhappy victims. These knives are known as *mandibles*, and are marked MD in the sketch, one on either side of the labrum. They first pierce the skin ; the *maxillæ*, marked MX, of which there are also a pair, then lap up the blood from the internal tissues. Finally, there is the true tongue or *lingua*, marked L, which is the organ for tasting it. As to the *maxillary palps*, marked MP, they do not form part of the tube at all, but stand outside it, and assist like hands in the work of manipulation.

This is how the mouth looks when fully opened out for microscopic examination. But as the fly uses it, it forms a closed tube, of which the labium and the labrum are the two walls, enfolding the lances or mandibles, and the lickers or maxillæ, as well as the tongue. Pack them all away mentally, from MX to MX, within the two covers, and you will then understand the nature of the mechanism. Look back at Fig. B in No. 9, and you will there observe that all the parts in the mosquito answer to those in the gadfly. The long upper sheath is the upper lip :

then come the lancets, the lappers, and the tongue, and last of all, the lower lip.

In No. 12, which is still more highly magnified, we have the essential parts of the blood-sucking apparatus made quite clear for us. Here LBR is the tip of the labrum, or upper lip, forming the front of the groove or sheath in which the lances work.

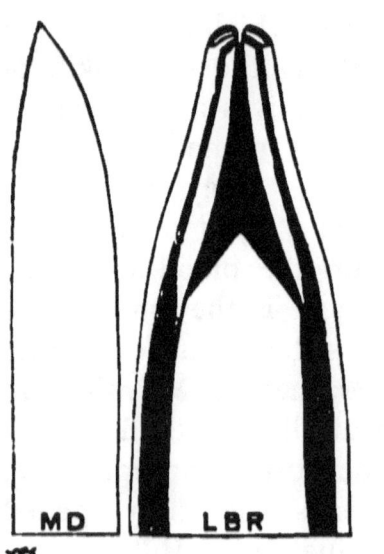

Its end is blunt, so as to enable it to be pressed close against the minute hole formed by the lances. MD is the sharp tip of one of the two lances, with its serrated or saw-like cutting-edge; this is the organ that does the serious work of imperceptibly piercing the skin and the tissues beneath it. MX

NO. 12.—THE CUTTING EDGES OF THE LANCETS.

is the tip of one of the maxillæ, or blood lappers, which suck or lap up the blood from the wound after the lances have opened it. I need hardly call your attention to the extraordinary delicacy and minuteness of these hard, sharp weapons, strong enough to pierce the tough hide of a horse, yet so small that if represented on the same scale as the insect itself, you would fail to perceive them.

Is it not marvellous, too, that the same set of organs about the mouth, which we saw employed by the wasp for cutting paper from wood, and by the ant for the varied functions of civilised ant-life, should be capable of modification in the butterfly into a sucker for honey, and in the gad-fly into a cunning mechanism for piercing thick hides and feeding on the life-blood of superior animals. Nature, it seems, is sparing of ground-plan, but strangely lavish of minor modifications. She will take a single set of organs, inherited from some early common ancestor, and keep them true in the main through infinite varieties ; but as habits alter in one species or another, she will adapt one of these sets to one piece of work and another to a second wholly unlike it. While she preserves throughout the similarity due to a common origin, she will vary infinitely the details and the minor structures so as to make them apply to the most diverse functions. Nothing shows this truth more beautifully, and more variously, than the mouths of insects ; and though the names by which we call the different parts are, I will admit, somewhat harsh and technical, I feel sure that anybody who once masters their meaning cannot fail to be delighted by the end-less modifications by which a few small instru-ments are made to fit an ever-increasing and infinite diversity of circumstances.

R

XI

A VERY INTELLIGENT PLANT

PEOPLE who have never had occasion to observe plants closely often fall into the error of regarding them as practically dead—dead, that is to say, in the sense of never doing or contriving anything active. They know, of course, that herbs and trees grow and increase; that they flower and fruit; that they put forth green leaves in spring and lose them again in autumn. But they picture all this as taking place without the knowledge or co-operation of the plant itself—they think of it as done *for* the tree or shrub rather than *by* it. Those, however, who have kept a close watch upon living green things in their native condition have generally learned by slow degrees to take quite a different view of plant morals and plant economy. They begin to find out in the course of their observations that the life of a herb is pretty much as the life of an animal in almost everything save one small particular. The plant, as a rule, is rooted to a single spot; the animal, as a rule, is free and locomotive.

Yet even this difference itself is not quite absolute: for there are on the one hand locomotive plants, such as that quaint microscopic vegetable tumbler, the floating green volvox, which whirls about quickly through the water like a living wheel, by means of its rapid vibratory hairs; and there are, on the other hand, fixed animals, such as the oyster and the sea-anemone, which are far more rigidly attached to one spot for life than, say, the common field-orchid or the yellow crocus. For field-orchids and crocuses do travel very slightly from place to place each season, by putting out fresh bulbs or tubers at the sides of the old ones, and springing up next year in a spot a few inches away from their last year's foothold; whereas the oyster and the sea-anemone settle down early in life on a particular rock, and never stir one step from it during their whole existence. Thus the distinction which seems to most people most fundamental as marking off plants from animals—the distinction of movement—turns out on examination to be purely fallacious. There are sedentary animals and moving plants; there are herbs that catch and eat insects, and there are insects that live a life more uneventful and more stagnant than that of any herb in a summer meadow.

Again, everybody who has studied plants in a broad spirit is well aware that each act of the plant's is just as truly purposive, as full of practical import, as any act of an animal's. If a child sees a cat lying in wait at a mouse's hole, it asks you why she does so; it is told, in reply, and truly told, " Because she wants to catch her prey for dinner." But even imaginative children seldom or never ask of a rose or a narcissus, "Why does it produce this notch on its petals? Why does it make this curious crown inside the cup of its flower?" Those things are thought of as purely ornamental; as parts of the plant, not as organs made by it. Yet the rose and the narcissus have just as much a reason of their own for everything they do and everything they make as the cat or the bird; they are just as much governed by ancestral wisdom, though the wisdom may in one case be conscious, in the other hereditary.

The rose, for example, produces prickles for its own defence, and scented blossoms to attract the fertilising insects for its own propagation. It does everything in life for some good and sufficient reason of its own, and takes as little heed of other people's convenience as the tiger or the snake does. "Each species for itself," is the rule of nature; no species ever undertakes anything for the sake of any other, except in the expectation of a corresponding advantage. If the wild thyme lays by in its throat abundant honey for the bumble-bee, that is because it counts upon the bumble-bee to carry its pollen from blossom to blossom; if the holly puts forth bright red berries for the robin to eat, that is not because it cares for the robin's distress, but because it looks upon the bird as a paid disperser of its stony seeds, and gives him in return a pittance of pulp for his pains, as stingy payment for the service rendered. The holly and the thyme are confirmed sweaters. Indeed, you will find that no plant ever wastes one drop more of nectar on its flowers, or one atom more of sweet pulp on its fruit, than is absolutely necessary to secure its own purely selfish object. It offers the bird or the insect the minimum wage for which bird or insect will consent to do the work it contracts for; and it never wastes one farthing's worth of useful material on tips or generosities. The rose, for all that poets have said of it, is strictly utilitarian. "You help me and I will help you," it says to the butterfly; and it keeps the sternest possible debtor-and-creditor account with all its benefactors.

As a familiar example of this purposive character in all plant life, I am going, in the present chapter, to take one of the most utilitarian shrubs—the common gorse—and try to show you why it behaves as it does in the conduct of its affairs; who help it in life and who hinder it, what friends it strives to buy or conciliate, what enemies it repels by what violent acts of armed hostility.

Everybody knows gorse; and everybody also knows that it is almost never out of flower. This last peculiarity, however, is due to a cause that not everybody has noticed. We have two distinct kinds of gorse at least—the larger and the smaller. It is the larger sort that one observes most when it is not in blossom, though it is the smaller kind whose golden bloom contrasts so beautifully in autumn with the rich purple of the upland heather. Now, the larger gorse begins to flower in October or November; it goes on opening its buds spasmodically in every fine spell throughout the winter, reaching its fullest glory of blossom in April and May; while the smaller kind begins to flower in summer, as soon as its larger cousin has fixed its attention on setting seed; and it goes on yellowing our heaths with its wealth of gold till October or November, when the bigger sort once more replaces it and takes up the running. In this way there is no bright day throughout the year— that is to say, no day fit for insects to gather honey—on which one kind of gorse or the other does not seek to cater for the friendly allies which help it to set its precious seeds, as we shall see in the sequel. It is the larger and better-known gorse with which I shall deal chiefly here, though I may occasionally refer by way of illustration or contrast to its smaller neighbour.

NO. 1.—THE BABY GORSE PLANT.

If we begin at the beginning in the life-history of the gorse, it may surprise you to find that each plant sets out on its way through life, not as a prickly gorse plant, but as a sort of quiet and un-armed little flat trefoil. No. 1 shows you the

young furze bush in its earliest infantile stage,
when it is still essentially a two-leaved seedling.
This seedling grows from a small bean scattered
by the parent plant in a very curious way, which
I will explain later. Thousands of the beans lie
on the ground in every common, and only a few
germinate, under favourable circumstances, into
two-leaved seedlings, like
those represented in these
illustrations. The leaves
of the first pair spread
out flat on the surface of
the unoccupied soil and
drink in the sunlight.
They also drink in, what
is equally important to
them, the carbonic acid
of the air, and manufac-
ture from it the living
material of fresh leaves
by the aid of the sun-
light. For the first few
days of its life, the
young gorse plant lives
mainly on the food laid
up for it in the bean by

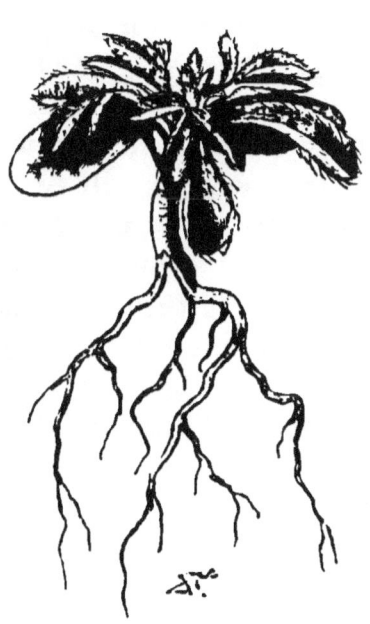

NO. 2.—THE GORSE PLANT AT
ONE WEEK OLD.

the parent bush ; but as soon as this is exhausted,
and it has accumulated a little stock of its own
by its private exertions, it begins to manufacture
new leaves and branches that it may rise above
the tangled mass of competitors by which its birth-
place is surrounded.

No. 2 shows us this second stage in the young shrub's development. At first sight you would hardly suppose it was a gorse at all; you might take it for the young of some such allied species as a broom or a genista. You will observe that at this point in its history the young gorse has trefoil leaves, not very unlike those of some kinds of clover. Why is this? Well, we have many good reasons for supposing that the ancestors of gorse were originally soft-leaved and unarmed shrubs, like the ornamental genistas which we grow in pots for drawing-room decoration; but as they were much exposed on open moors and commons, where they were liable to be grazed down and browsed upon by rabbits, sheep, and other herbivorous animals, the tenderer and more luscious among them stood little chance of surviving. Indeed, so hard is it for plants to grow in such situations, that one not uncommonly finds tiny trees of Scotch fir, close cropped to the ground, yet with many years' growth exhibited by the annual rings of wood in their underground root-stock. These poor persistent little trees have been nibbled down, year after year, as soon as they appeared, by rabbits or donkeys; yet year after year they have gone on sprouting afresh, as well as they could, and laying by an annual ring of woody tissue in buried root-stock.

To some such attacks the ancestral gorses must always have been exposed on the open moors and hillsides of primitive Europe, at first, no doubt, from deer and wild oxen and beavers, but later on from the sheep and cows and goats and donkeys which followed in the wake of aggressive civilisation. Under these circumstances, most of the soft-leaved and unprotected plants got eaten down and killed off; but any shrub which showed a nascent tendency to develop stout spines or prickles on their branches must have been favoured by nature in the struggle for existence. The consequence was that in the end our upland slopes and open spaces all over Western Europe came to be occupied by nothing but strongly armed plants — brambles, thistles, blackthorns, may-bushes, nettles, butcher's-broom, and the various kinds of furze, all of which can hold their own with ease against the attacks of quadrupeds. Indeed, there is one not uncommon English herb, the little purple-flowered rest-harrow, which very well illustrates this curious connection between the production of thorns and the habit of growing in much - browsed-over spots; for when it settles in enclosed and protected fields it produces smooth and unarmed creeping branches, but when it happens to find its lot cast in places where donkeys and rabbits abound, it defends itself against the dreaded enemy by covering its shoots with stout woody prickles.

Still, to the end of its days, the developed gorse plant never entirely forgets that it is the remote descendant of trefoil-bearing ancestors; for not only does every young gorse begin life with trefoil foliage, but if frost happens to check the growth of the budding branches in the full-grown bush, or if fire singes them, the shrub at once puts forth a short sprout of trefoil leaves at the injured point, as though reverting in its trouble to its infantile nature.

NO. 3.—THE PLANT OUTGROWING
ITS TREFOIL STAGE.

In No. 3 we see the third stage in the upward evolution of the baby gorse. Here, the seedling begins to outgrow its childish trefoil stage, and to prepare itself for the repellent prickliness of its armed manhood. You will observe in this case that the outer and lower leaves have still three leaflets apiece, but that the upper and inner ones —that is to say, the youngest and latest produced — have the form of single long blades, like those of the broom bush. As yet, these solitary leaves are also unarmed : they do not end in sharp points like the later foliage, and they cannot pierce or wound the tender noses of

sheep or rabbits. But if the gorse were to continue long in this unarmed condition, it would
stand a poor chance in
life on its open hillsides;
so it soon proceeds to
the stage exhibited in
No. 4. This illustration
shows you a plant about
a fortnight or three
weeks old, with trefoil
leaves below, passing
gradually into silky and
hairy single blades,
which in turn grow
sharper and thinner as
they push upward to-
wards the unoccupied
space above their native
thicket. Interspersed
among these sharp little
leaves you will also note
a few grooved branches,
each ending in a stout
prickly point; these
prickles are the chief defence of the bush against
its watchful enemies.
But the leaves and the
branches are often so
much alike that only a

NO. 4.—THE YOUNG SHRUB BEGINS
TO ARM ITSELF.

skilled botanist can distinguish the one from the
other. Both are sharp and intended for defence;

and as the branches of gorse are green like the leaves, both perform the same feeding function.

In No. 5 I have chosen for illustration and comparison a full-grown shoot of the common-

NO. 5.—ITS FIRST COUSIN, THE GENISTA.

scented yellow genista, so often grown in pots as a table decoration. This pretty shrub begins in life so much like a gorse-bush, that if I were to show you very youthful seedlings of both, you could hardly discriminate them. That is to say, in all probability, both are descendants of a common ancestor which had trefoil leaves and bright yellow peaflowers. But the scented genistas happened to find their lot cast in inaccessible places, on cliffs or crags, where defence against browsing animals was practically unnecessary; while our ruder northern gorse had its lines laid on rough upland moors, where every passing beast

could take a casual bite at it. The gorse was,
therefore, driven perforce into producing thorny
branches which would repel its foes, while the
genista retained the old soft silky shoots and
broad trefoil foliage.
Broom, which is a close
relation of both these
plants, with much the
same yellow peaflowers
and hairy pods, occu-
pies to some extent an
intermediate position
between the two types.
The young shoots have
leaves of three leaflets,
as shown in No. 6 ; but
the older branches are
covered with leaves of
a single leaflet apiece,
like the second form
produced by the gorse
plant. The trefoil leaves
of the broom also
closely resemble those
of the laburnum, which
is another and more
tree-like descendant of
the same ancient an-

NO. 6.—ITS SECOND COUSIN,
THE BROOM.

cestor, with similar yellow blossoms, and pods and
beans of much the same character. It is interesting to observe in a family of this sort how the
young seedlings are in every case almost identical,

and how, as they approach maturity, they begin to assume the adult differences which mark off each later developed kind from the primitive and central form of its ancestors.

But is gorse really exposed to the attacks of animals ?   Would any herbivore care to eat such hard food ?   If you doubt it, you have never lived near a gorse-clad common.   From the moment the seedling shows itself above the ground it is cease-lessly nibbled at by rabbits and other rodents ; and even after it has acquired its prickly armour, it makes excellent fodder, if only the sharp tops can be rendered harmless to the

NO. 7.—PROTECTING THE BUDS FROM
BROWSING ANIMALS.

sensitive noses of cattle or donkeys.   Gipsies know this fact well ; and you may often see them on our Surrey hills cutting the succulent young branches and chopping them up fine in a wooden trough till the prickles are destroyed.   Their horses then eat the good green food most greedily.

The gorse knows the same thing, too ; and it takes particular care to preserve its leaves and flowers against the aggressive quadrupeds. When November comes it begins to blossom. No. 7 shows you how cleverly and cautiously it makes its preparations for this important function. The flower-buds, I need hardly say, are particularly rich and juicy, and, there-fore, particularly liable to the assaults of the enemy. Hence, you will observe, they are doubly protected. To guard against large animals, each little knot of buds is care-fully placed, for safety, in the angle formed by the main stem with one of its short, stout branches. Stem and branch alike end in a forbidding prickle, and the buds are so set in the axil that

NO. 8.—THE GREAT-COAT, PROTECTING THE BUDS FROM COLD AND FROM EGG-LAYING INSECTS.

it is simply impossible for any browsing creature to get at them without encountering both these serious weapons. Indeed, no illustration can fully bring out the beautiful variety and complexity of arrangement by which each separate group of buds is completely defended ; in order to understand it fully, I advise you, after reading this chapter, to go out to the nearest common, and examine a flowering gorse-bush for yourself, when you will see how wonderfully and how intelligently the plant provides for the equal security of all its blossoms. I do not wish to be personal, but if for one moment you can imagine yourself a donkey, and try to help yourself with your teeth to some of the juicy buds, you will find that it is practically impossible to do so without receiving a whole array of serried lance-thrusts from several separate prickles.

But large animals are not the only foes against which the gorse has to defend its blossoms. It is almost equally exposed to the unfriendly attentions of flying insects, which desire to lay their eggs near its rich store of pollen and its soft yellow petals. To ward off these winged assailants, mere prickles are insufficient. The insect can wriggle in sideways, and so deposit its egg, which would develop in time into a hungry grub; the grub would proceed to eat up the flower, and thus defeat the object which the plant has in view in producing its blossoms. No. 8 shows you how the gorse meets this second difficulty. It covers up the buds with its stout calyx, which, for greater security, is reduced to a pair of sepals only, though in allied types there are five, and traces of the five still exist in the lobed top of the existing calyx. This outer coverlet, or greatcoat, is thickly sprinkled with a sort of fur, composed of dark brown hairs, which baffle the insects, and prevent them from laying their eggs upon the surface. Indeed, nothing keeps off insects so well as hairs; they form to these little creeping creatures an impenetrable thicket, like tropical jungle to an invading army. Ants, you will remember, cannot creep up stems which are thickly set with hairs; and in warm climates, people take advantage of this peculiarity by wrapping fur round the legs of meat-safes, so as to keep off those indefatigable pests of the equatorial housekeeper.

Nor is this the only use of the short brown hairs. I spoke of the calyx above as a great-coat, for warmth is really one of its chief objects. It keeps off the cold as well as the insects. You must remember that the greater gorse is a winter-flowering plant: it lays itself out to attract the few stray bees which flit out in search of food on sunny mornings in December and January. A bush with this habit needs protection for its buds from the cold: just as you see the crocus does, when it wraps up its flowers in a papery spathe, and as the willow does when it encloses its catkins in soft, silky coverings. The hairy coat of the gorse-bud has just the same function: it is there for warmth as well as for protection against egg-laying insects. That, I think, is the reason why the hairs are coloured brown; because brown is a good absorber of heat; the fur collects and retains whatever warmth it can get from the winter sun in his friendlier moments.

You will further observe in the illustrations, and still better on the living gorse-bush, that all the buds are not at the same stage of development

s

together. The plant does that intentionally. It is
a slow and gradual flowerer. The reason is plain.
Our winter and spring are proverbially uncertain.
The bush does not want to put all its eggs into one
basket. Sometimes, in doubtful weather, a few
of the buds develop up to the stage shown in
No. 8, and are just ready to open. Then comes
a frost, a killing frost, and nips them in the bud,
more literally than we often mean when we use
that familiar metaphor. In such cases, you will
sometimes find the more advanced flowers are
killed off and never develop further. But look
behind them in No. 8, and you will see that the
bush holds in reserve a number of younger buds,
against this very contingency. They are wrapped
up tight in their warm brown overcoats, and they
keep one another warm as they nestle against the
stem ; so that however sharp the frost, they seldom
suffer, in England at any rate. Beyond the Rhine,
where the winters are severer, both buds and foliage
would be nipped by the east wind ; and so the
smaller gorse is confined to the portion of Europe
west of the Rhineland, while even the greater kind
cannot live in Russia. To eastward its place is
taken by hardier shrubs, which have still more
special methods of protection against the severe
weather. In Western Europe, on the other hand,
the buds are so arranged that in spite of frost we
get a constant succession of gorse-blossoms from
November to May or June, when the running is
taken up by the smaller summer species. Thus
the bees are never deprived of gorse-blossom,

and kissing, as the old saw says, is never out of fashion.

I have said above that gorse protects itself against flying insects. But not indiscriminately. It is a respecter of persons. While it wishes to keep off the egg-laying and flower-gnawing types, it wishes to attract and allure the honey-suckers and fertilisers. For this object alone it produces its bright yellow petals and its delicious, nutty perfume, which hangs so sweetly on the air in warm April weather. And I know few things in plant life more instructive and interesting to observe than the way of a bee with this flower. Go out and watch it, and verify my statements. When the blossom first opens, it looks somewhat as in No. 9, only that the keel, as we call the lower part of the flower, is not half open, as there, but firmly locked together above the stamens on its upper edges. This keel, as you may note in No. 10, consists of two petals slightly joined together at the margin. On either side of it come two other petals, which we call the wings, and which are fitted with a funny little protuberance at their base so arranged that it locks the whole lower part of the blossom together. This mechanism cannot be seen in the illustrations, nor indeed can it be properly understood except in action; but gorse is so universal a plant that most of my readers can observe it and examine it for themselves at leisure. The upper petal of all, known as the standard, has no special duty to perform save that of advertisement. It attracts the insects, and shows them in which direction to approach the flower.

Now comes the strangest part of the whole process of flowering. When the bee settles on the blossom, she alights on the keel and wings, to which she clings by her fore-legs, and so weighs down the entire lower portion of the mechanism with her weight. As she does so, the clasps or knobs on the wings come undone, and the whole flower springs open elastically, as you see it in No. 10, exposing the stamens and the young pod which form its central organs. At the same moment, the pollen, which is specially arranged for this contingency, bursts forth in a little explosive cloud, covering the body and legs of the visiting insect. She takes no notice of this queer manœuvre on the part of the plant, being quite familiar with it, but goes on helping herself to the store of honey. As soon as she has rifled it all, she flies away, and visits a second flower of the same kind. In the act of doing so, she rubs off on its sensitive surface the pollen with which the

NO. 9.—THE FLOWER, HALF OPENED.

last blossom dusted her, each part being so con-
trived that what she takes from one flower she
hands on to another. You can see the little tufted
stigma standing up in the centre of No. 10, and can
understand how it must catch on its tip the fertilising yellow grains which the bee collected in a previous explosion.

But now no-
tice a curious
thing that next
happens. When
once the flower
is "sprung," as
we call it—that
is to say, thus ela-
stically opened
—the keel and
wings never go
back again into
their original
position. They
remain perma-
nently open.
You will thus
comprehend that
there is a great
difference between the virgin flower, in which the
keel and wings are locked over the stamens, and
the "sprung" one, in which the keel and wings
have descended from their first position so that
the entire centre of the blossom is exposed to
view. Moreover, after the flower is once ferti-

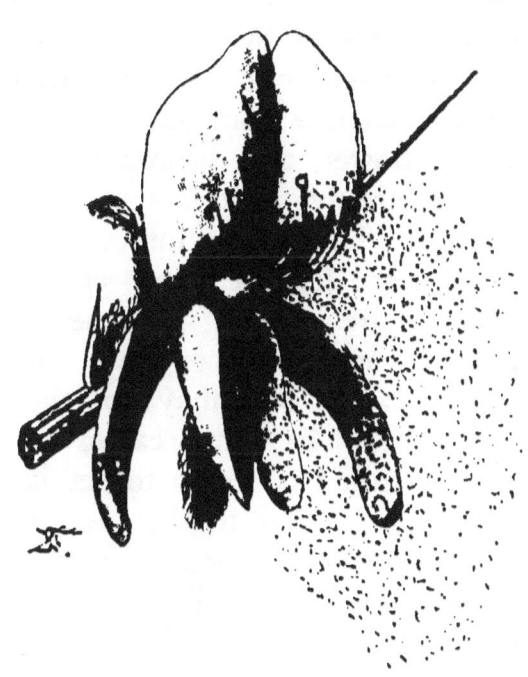

NO. 10.—THE FLOWER, SPRUNG, AND
DISCHARGING POLLEN-SHOWERS.

lised, it produces no more bribes for the bee; it has got all it wants out of her, and it is certainly not going to find her in food and pay her wages for nothing. The consequence is, that a "sprung" flower becomes, as it were, an advertisement to the bee of "Nothing to eat here." If you watch a bee paying her visits to a gorse-bush, you will find that she passes by the "sprung" flowers without the slightest notice—seems, in fact, oblivious of their existence; but she fastens at once on each virgin flower, and promptly—though, of course, unconsciously—fertilises it. Such a device for showing the visiting insects automatically which flowers are fertilised and which are not is, naturally, a great saving of time; and plants which develop such devices gain such an advantage thereby as neither they nor the bees are slow to appreciate. In some cases, as seen, as soon as the blossom has begun to set its seeds, it changes colour as a sign to the bees and butterflies that it is no longer open to receive their visits; in others, the petals fall the moment fertilisation is effected, and so the flower ceases to be at all conspicuous.

In the gorse-bush, the petals, however, do not fall at all. They remain to enclose the young pod as it swells and develops. The reason for this divergence from the usual habit of plants is, I think, because the gorse-bush flowers and ripens its fruit in such very cold weather, that the young and tender pods need all the cover they can get at the moment when they begin to swell and to go through the important process of fructification. The calyx and the petals help to keep things warm for them, and so they persist till the pods are ready to open and discharge their beans.

Each pod contains as a rule four beans, and these are fat and well stored with nutriment for the baby seedling. The young plant subsists for its first few days on the nourishment thus laid by for it ; for gorse is not one of those improvident plants which turn their young ones loose upon a cold and unsympathetic world without a coin in their pockets, so to speak, to fall back upon. Plants in this respect differ, like human beings. Some send their offspring out, mere street arabs of the vegetable world, without any capital to live upon ; others provide them with a good stock or reserve of foodstuff which suffices them till they are of an age to earn their own living. You can judge by the fatness and distention of the pod in No. 11 that the young beans of the gorse are fairly provided for in this respect. Indeed, so rich are they in food, that they would suffer seriously from two sets of enemies, were they not protected against both exactly as the buds are. The stout prickles at the ends of the branches efficiently repel the assaults of browsing animals ; the close hairs on the pods (not seen in the sketch) just as efficiently repel the insects which would fain lay their eggs in the beans, as one knows they do in the similar case of the edible peas in our garden.

Nothing is more beautiful about the gorse,

indeed, than the soft, close covering of fur in the young pods, which gives them almost the appearance of miniature ducklings. No insect can penetrate it; and if only the first few days pass by without serious mishap, the gorse may count upon maturing its seeds in peace and quietness.

They ripen in the first basking warmth of July, or often earlier. As soon as they are ready for dispersal, the bush has a device for scattering them and sowing them in proper places for their due germination, which is quite in accordance with its other proceedings. Gorse, indeed, is a very explosive species. It knows the full value of the propulsive habit. The valves of the pods remain straight and rigid after the beans have ripened; but the sides contract, only the ribs or thickened edges keeping them extended in their places. At last, on some very sunny morning, the baking heat dries them up to such a point that they can no longer hold together. They curl up suddenly and violently, as you see in No. 12, and expel the beans, shooting them out like little bullets all over the common. If you happen to sun yourself on a gorse-clad

NO. 11.—THE POD, WITH THE BEANS WITHIN IT.

moor on such a warm summer morning, you will hear, from time to time, little abrupt discharges as if a succession of toy pistols were being continually fired off in the thicket all round you. These noises are due to the bursting pods of gorse, which go off one after another, and shed their seeds piecemeal over a considerable area.

Should you look in early spring on the bare spots around a moor or common, you will find gorse seedlings by the thousand, all fighting it out among themselves, and all trying their best to occupy the uncovered

NO. 12.—THE POD, AFTER DISCHARGING THE BEANS ELASTICALLY.

spaces in the neighbourhood of their parents.

And here the wonder of their lives begins all over again. For while the gorse was old and woody, it grew like gorse, all stern and prickly. But as soon as the young seedlings start afresh in life, they seem to forget their parents : they revert once more to the old trefoil condition. All young plants and animals, at least in their embryonic stages, show this strange tendency to throw back

at first to the ancestral form; and it is fortunate
for us that they do so, for it often enables us to
perceive underlying relationships which in the
adult form escape our notice. Nobody who looked
at a furze-bush in its stiff and prickly old age
would ever suspect it at first sight of a cousinship
with clover. Yet when we consider the trefoil
leaves of the seedling, and the shape of the separate peaflowers in the adult form, we can see for
ourselves that the two plants are far closer together
than we might be tempted to imagine. Indeed
between the little creeping yellow clovers and the
aggressive furze or the tall and beautiful laburnum,
we can find even now a regular series of connecting links which show clearly that all alike are
slightly divergent descendants of a single common
ancestor.

We may conclude, then, that gorse in every
particular lays itself out in life to fight its own
battle, and to meet the peculiarities of its special
situation by its own exertions. Born a trefoil-
bearing plant, unarmed and undefended, it produces spines instead of leaves as soon as its growth
exposes it to the attacks of enemies. It defends
its buds alike from the attacks of cattle and the
assaults of insects; it wraps them up from the cold
in efficient overcoats. It cares for its young and
lays up food in its beans on their account; it
scatters its seed upon unoccupied spots where they
may stand the best chance of picking up a living.
All these acts are analogous to those produced by
intelligence in animals; and though the intelligence

is here no doubt unconscious and inherited, I think we are justified in applying the same word in both cases to operations whose effects are so closely similar. Gorse, in short, may fairly be called a clever and successful plant, just as the bee may be called a clever and successful insect, because it works out its own way through life with such conspicuous wisdom.

XII

A FOREIGN INVASION OF ENGLAND

OUR worst enemies are not always the most apparent ones. It is easy enough to build forts for the protection of our towns and harbours against French or Germans, but it is very difficult to devise means of defence against such insidious foreign invaders as the influenza germ or the Colerado beetle. France lost much by the war with Germany, but she probably lost more by the silent onslaught of the tiny phylloxera, which attacked her vineyards—attacked them, literally, root and branch, and paralysed for several years one of her richest industries. Yet invasions like these, being less obvious to the eye than the landing of a boat-load of French or German marines on some bare rock in the Pacific claimed by Britain, attract far less attention than aggressions on the Niger or advances in Central Africa. The smallness of the foe makes us overlook its real strength—it has the force of numbers. We forget that while we can exterminate hostile human bands with Armstrongs and torpedo-boats, the resources of civilisation are still all but powerless against the potato blight, the vine disease, and the destroying microbe.

The enemies of our corn crops in particular are many and various. There is the wheat-beetle, for example, which ravages the wheat-fields in two ways at once, the grub devouring the growing young leaves, while the perfect winged insect eats up at leisure the grain as it ripens. There are the various cockchafers, which vie with one another in their cruel depredations on the standing corn. There are the skip-jacks and wire-worms, and other queerly named beasties, which attack the roots of the plant underground. There is the corn saw-fly, whose larva feeds on the stalk of rye and wheat, till it finally cuts off the whole haulm altogether close to the soil at the bottom. There are the midges which lay their eggs in the swelling ear, where the maggots develop and prevent the proper growth of the impregnated grain. There is the gout-fly, which causes a gouty swelling at the joints, and the corn - moth, which devours the stored wheat in the granary. There are the red-maggot, and the grain-aphis, and the thrips, and the daddy-longlegs, all of which in various ways prove themselves serious enemies of the agricultural interest. And there are dozens more known only to men of science by dry Latin names, and duly chronicled by the farmer's friend, Miss Ormerod, in many learned and exhaustive monographs.

But as if these were not enough for our "depressed" neighbours, the agriculturists, the last ten years or so have seen England invaded by a foreign foe, either from Germany or America— a foe whose life-history has been made a special subject of study by my collaborator, Mr. Enock, and whose strange story I shall detail (largely from his materials) with no unnecessary scientific verbiage in this present chapter.

The new invader is called the Hessian fly; and he made his first appearance in Britain, or at least first attracted official entomological attention in this country, in 1886. If he was here earlier, he skulked incognito. For more than a century, however, he had already been a great scourge in America, where he first acquired the name of Hessian fly during the revolutionary war, through the popular belief that he had been imported from Europe into Pennsylvania by the Hessian troops employed as mercenaries by George III. in his fruitless struggle against the revolted colonies. The Hessians were the *bêtes noires* of the patriotic Americans; and the farmers, finding their crops devastated by a pest till then unknown, came at once to the conclusion that their enemy, King George, had sent the two plagues, human and entomological, over sea together. They regarded the question much in the same spirit as that of the loyal poet in the " Rejected Addresses," when he asks about Napoleon, " Who fills the butchers' shops with large blue flies ? " The Briton set down every natural misfortune to " the Corsican ogre "; the American set down all evils that befell him to the Rhenish mercenaries.

Ever since that day, much controversy has raged in America and Germany as to the original home of the destructive creature. One school of dis-

putants hotly maintains that the Hessian fly, which
now abounds in parts of France, Austria, and
Russia, is a native of the Old World, and that its
first home coincided with that of our primitive
cereals, Southern Europe and Western Asia.   Another school, anxious to make out the enemy an
American citizen, fights hard for its being an
aboriginal inhabitant of the United States.   Thus
much, at least, is certain, that at the present day
the "fly" is found in both hemispheres in too
great abundance, and that in America in particular
in certain disastrous years it has almost ruined
the entire wheat crop.   I have seen whole fields
upon fields there simply pillaged by its ravages.
The loss produced by this insignificant little creature, indeed, has in some seasons been measured
by millions of pounds sterling.

If you go out into a barley-field in England
where the Hessian fly has effected his entrance,
you will probably find a large number of plants of
barley, like that delineated in No. 1, with the
stem bent down sharply toward the ground at the
second joint.   At first sight you might imagine
these stalks were merely broken by the wind or
fallen by their own weight; but if you examine them closely in the neighbourhood of the
bend, which occurs with singular unanimity in
all the affected plants at about the same point,
you will find inside the sheath of the blade, where
it encircles the stem, a curious little body which
the farmers with rough eloquence have agreed
to describe as a "flax-seed."   If you watch the

development of the "flax-seed" again, you will
find that it is not a seed at all, but the pupa-
case (or rather the grub-shell) of a small winged
insect; and it is the life-history of this insect, the
Hessian fly, that I now propose to sketch for you
in brief outline.

No. 2 shows the mother fly herself, very much
enlarged, for in nature she is but a small black
gnat, belonging to the same group as our old

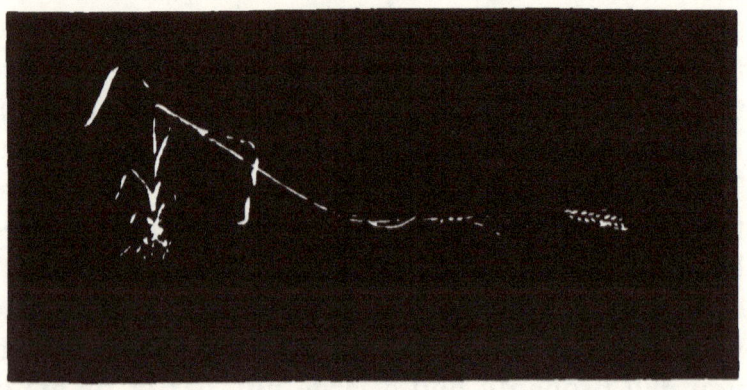

NO. I.—AN INVALID BARLEY PLANT.

friend (and foe) the mosquito. You will observe
that she is a fairy-like creature, for all her wickedness: she has two delicately fringed wings (with
"poisers" behind them), a pair of long antennæ
with beaded joints, six spindle legs, and a very
full and swollen body. She needs that swollen
body, for she is a mighty egg-layer. She flies
about on the stubbles in September, and lays her
eggs on the self-sown barley plants and on the
aftergrowth of the cut crops: as well as in spring

(a second brood) on the new sprouting barley. One industrious female which Mr. Enock watched when so employed laid no less than 158 eggs on six distinct plants ; while another laid eighty on a single leaf.   He has noted in detail many cases in the same way, and all show an astonishingly high level of maturity.   The eggs are extremely minute, and are pale orange in colour, with reddish dots. Most of them are deposited on the leaf itself, or on the sheath or tube which forms its lower portion.

And now see how clever this dainty little creature is !   She lays

NO. 2.—THE SOURCE OF THE MISCHIEF : THE HESSIAN FLY.

her eggs with the head end downward ; and as soon as the tiny grub hatches, which it does about the fourth day, it emerges from the shell, and walks straight down towards the stem, at the point where the protecting leaf-sheath is wrapped closely round it.   The worm forces itself in between the stem and the sheath, and after

T

walking steadily for four hours, at the end of
which time it has covered a record space of
nearly three inches, it arrives at the joint, where
the sheath begins, and so finds its way blocked by
the partition wall; it can get no further. Here

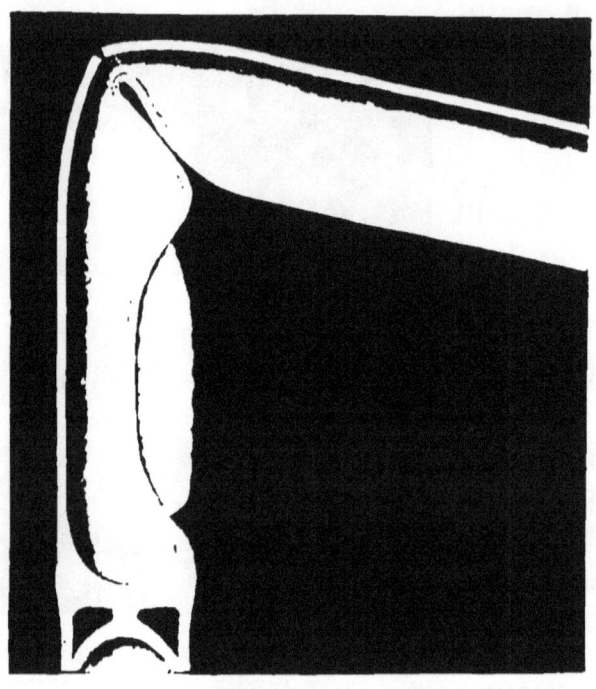

NO. 3.—THE GRUB AT WORK.

then the young grub stops, as you see in No. 3,
wedged tightly in between the leaf-sheath and the
stem, and with its head pointing downward. Being
a hungry, and therefore an industrious creature,
it at once sets to work to eat the barley-plant.
This it does by fixing its sucker-like mouth on the

soft, sweet, and juicy portion of the stem just above
the joint—that same soft, sweet, and juicy portion
which children love to pull out and suck, and
from which the grub, too, sucks the life-juice of the
barley-plant.   Naturally, however, you can't suck
a plant's life-blood without injuring its growth ; so,
after a very short time, the enfeebled stem begins to
bend, as you see in No. 3, a little distance above
the point where the grub is devouring it.   It has
been undermined, and its vitality sapped, so it gives
way at once near the source of the injury.

How much damage this action does to the crop
you can best understand by a glance at the two
next contrasted illustrations.   No. 4 represents
"seven well-favoured ears" of barley, unaffected
by Hessian fly, and with the grains richly filled out
as the farmer desires them ; No. 5, on the contrary,
shows you "seven lean ears," attacked by the fly,
and bent and ruined in various degrees by the
indirect action of the silently gnawing larva.   Look
on this picture and on that, and you will then
appreciate the British farmer's horror of his insignificant opponent.   You will observe, by the
way, that I speak throughout of barley, not of
wheat.   This is because in England, where these
sketches are studied, the time of wheat-sowing is
such that the wheat has so far escaped the pest ;
the female flies are all dead before the crop is
sprouted : whereas in America the "fall wheat"
comes up at the exact moment when the female
Hessian fly is abroad and scouring the fields in
search of plants on which to lay the eggs of her

future generations. In England, therefore, it is
barley alone which is largely attacked; and since
barley is mainly used for malting, to make beer
or whisky, the teetotaler may perhaps reflect with

NO. 4.   SEVEN WELL-FAVOURED EARS, UNATTACKED.

complacency that the fly is merely playing the
game of the United Kingdom Temperance Alliance.
His joy, however, is fallacious, for, on the other
hand, if we don't raise enough barley at home to
brew our ale, we don't on that account refrain

from malt liquors : we buy it from elsewhere ; so that, in the eyes of the impartial political economist at least, the Hessian fly in Britain must be regarded as an unmitigated national misfortune.

The grub eats and eats, in his safe cradle between the sheath and the stem, till he is ready to pass

NO. 5.—SEVEN LEAN EARS, ATTACKED BY GRUBS.

into the adult condition. But he does this by various and complicated stages, all of which I do not propose to set forth in full with the tedious minuteness of a scientific treatise, lest I weary that fastidious and somewhat lazy person, the "general reader." It must suffice here to say, in brief, that the larva is at first soft and free, but that before becoming a true pupa or chrysalis he passes through an intermediate encased or "flax-seed" stage, in which he performs some curious evolutions. The young larva when he starts in life is whitish or yellowish; in the "flax-seed" stage he becomes a rich chestnut brown, and seems externally quiescent. But the fact is, he arrives at full growth in the white form, and then leaves off feeding; his skin now hardens and darkens, and he looks from outside very much like a pupa. Indeed, his outer covering is now a sort of solid pupa-case, in shape just the same as the original grub, but more sombre in colour. No. 6 shows you the portrait of the grub in this curious intermediate condition. If you compare it with No. 3, you will see that the outer skin still preserves the original shape of the fat young larva; but that the enclosed grub himself, here shown as if the case were transparent, has shrunk away from his own old skin, just as a ripe nut shrinks away from its shell, to borrow Mr. Enock's admirable phrase for describing the process. And this strange shrinkage is connected with a very curious fact in the eventful life-history of the Hessian fly; it tells us of a problem which the grub has to face, and for which it has devised a most unexpected solution.

You remember that the young maggot had necessarily to work its way *head downward* along the stalk, in order to fix itself in the only place where it can find the soft food needful for it, between the sheath and the stem, where the tissue

is tenderest. But when it emerges later on in the
open air as a fly, it has to walk back again to the
outer world above the joint; and this it could not
do if it had still to go head downward. Yet there
seems no room for it to turn in. Somehow or

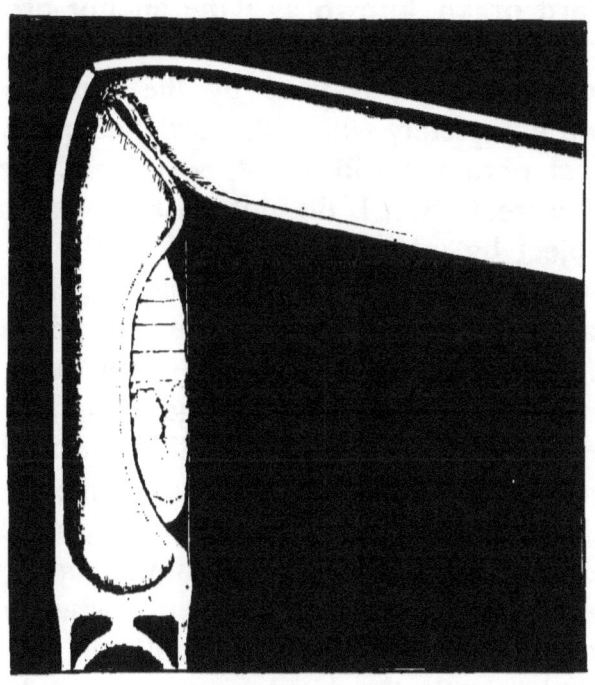

NO. 6.—THE GRUB TURNING ROUND INSIDE ITS OWN SKIN.

other, in that restricted space, it must reverse its
position; it must get itself head upward. How
is it to do so? This difficulty early struck Mr.
Enock in his examination of the creature's life;
and with characteristic patience he determined to
investigate it. His researches not only answered

the question itself, but also discovered a meaning and purpose in a certain organ of the adult grub, the nature of which had heretofore been a standing puzzle to that section of society which interests itself prominently in the Hessian fly question. The larva in its "flax-seed" stage develops an odd and very hard organ, known as "the anchor-process," near the head; and this "anchor-process," as Mr. Enock has shown, is used by the grub to turn it round completely within its hardened pupa-case. (The last phrase, I will admit, is not quite scientifically correct, but I do not wish to complicate the subject by introducing a multiplicity of technical terms unknown to my readers.) In No. 6 you can see the adult grub in the very act of thus turning round, head to tail, within his outer skin, so that he may be able to emerge as a full-grown fly, head upward. A tiger is nothing to it, though a tiger moves within his own integuments more freely than most of us. You will note that during the feeding stage the grub's mouth and under side were pressed against the stem; when he has performed this curious somersault on his own axis, so to speak, the head is uppermost, but the mouth and under side of the body are turned outward towards the sheath, not inward towards the stem and hollow centre of the barley-plant. He wants now to bite his way out, not to suck at the stalk for its nutritive juices.

I need hardly add that it takes some watching to detect such invisible movements inside a hard dark case; and only by the closest and most unwearying attention was Mr. Enock enabled to discover the true use and meaning of the so-called " anchor-process." It is really not an anchor, but a sort of hooked foot or lever, by whose aid the apparently dormant grub turns himself bodily over within his own hardened skin, now become too large for his shrunken body.

Discoveries like these are hard to make ; yet they bring little return in money or glory. But it is only by such patient and careful investigation that a way can be discovered to get rid of pests which cost civilisation many hundreds of thousands, nay, many millions, annually.

The grub in the turning stage is thus by no means what he looks—a dormant creature ; on the contrary, he is a gymnast of no small skill and activity. The muscular contortions by which he seeks to free himself of discomfort when disturbed by man show that he possesses great power of contraction, and that he can exercise a considerable force of leverage.

After the grub has succeeded in putting itself in position for assuming the winged stage, and emerging from its home head upward, it begins next to grow into a true pupa, or chrysalis. It is in the pupa, of course, that all winged insects acquire their wings and become definitely male or female, and this stage is, therefore, one of the most important. As soon as the grub begins to reach it, he swells once more and grows quite tight inside his larval skin, which is stretched so much that it seems to be bursting. At last, as he wriggles and twists within it, the skin does burst, first over the mouth and head, and then over the central joints of the body. Again the insect twists and wriggles inside this half-broken skin, and again he pushes it backward toward his tail, till at last he has sloughed it all off entirely, and it remains a shrivelled relic—an empty case—in the spot where he has hitherto lived and breathed and had his being. He is now a true pupa, white at first, but gradually growing a delicate pink, and then rosy.

Just at first, however, the pupa looks almost as formless as the grub it replaces, revealing no limbs or distinct segments. But little by little, feet and legs and eyes and wings begin to be visible through the semi-transparent shell of the chrysalis. He is changing slowly into a winged insect, and you can watch the change through the delicate horny coverings.

Stranger still, the Hessian fly at this stage is not torpid and quiescent like most ordinary insects. The pupa, as in many of this family, is locomotive. It has legs and feet, and it can wriggle its way up, as you see in No. 7, where the lower object is the empty larval skin, now deserted by its inmate, while the upper one is the pupa, emerging from the sheath, and making its first experiences of the wide, wide world outside its native leaf-bound hollow. It is ready now to come forth from the pupa stage, and to fly forth in the open air in search of a mate with whom to carry on the serious business of replenishing the fields with new generations of similar larvæ.

The succeeding illustrations show you in detail the various stages in the process of emergence. No. 8 gives you the beginning of emancipation. The pupa has here bitten its way through the leaf-sheath with its hard, horny jaws, and is pro-truding visibly. Just at first, only the head itself gets free; then the in-sect rests a while after its ardous labour, and begins wriggling and writhing again, this time working out its body or thorax. After another short interval for recuperation after such a terrific effort, it manages to pull its legs through the hole, and to support itself upon them by resting them like a bracket against the stem of the barley. This is the point just reached in the illustration No. 8. There the pupa stops short, having got himself

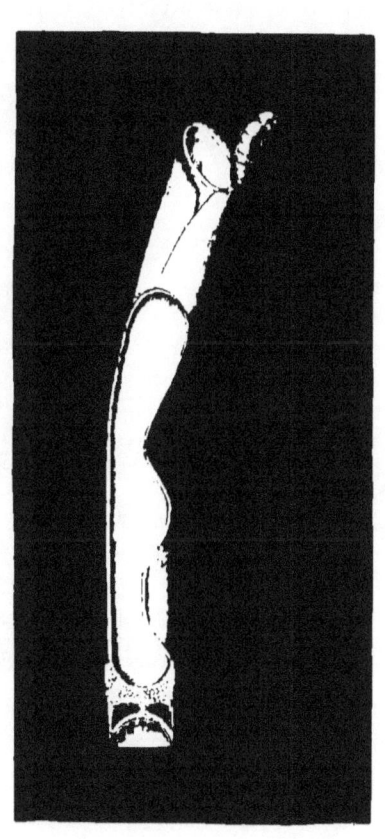

NO. 7.—THE CLIMBING PUPA; BELOW, THE EMPTY CASE.

into a convenient position for dispensing with his coverlet; for the sheath of the barley grasps the pupa-skin tight as in a vice, and he can wriggle his winged body free within it, without paying any further attention to the disused mummy-case which once confined it.

In No. 9, the pupa being thus safely anchored, the fly is emerging. It is a slow and delicate process, for with so many legs and wings and

NO. 8.—THE PUPA COMES OUT.

antennæ and appendages to get free from the mummy-case, one cannot hurry; haste might be fatal. At this first stage of emergence, as you will observe, all the important parts are still cramped at their ends within the pupa-shell; but you can see how the legs and antennæ are striving to dis-

engage themselves.   The pupa covering is propped
as before by the empty leg-shells so as to form a
bracket.

In No. 10—hurrah! with a supreme effort, our
fly has got her antennæ free! She can move them

NO. 9.—AND THE FLY COMES OUT OF IT.

to and fro now, in all their jointed and tufted glory.
That enables her to wag her head in either direction without difficulty, and encourages her to go
on to fresh exertions for the rest of the deliverance.
But her feet are still fast in that hampering mummy-
case ; she must try her hardest now to free them
each carefully.

First, however, let her get the tips of her wings
free to help them. One good jerk and out comes
the first wing. Now she bends backward and
forward and seems straining every nerve. Halloa,
that did it; the other wing is free! Not as yet,

NO. 10.--ANTENNÆ FREE !

however, plimmed out and flattened as it will be a
little later; both wings at present look somewhat
thick and lumpy and stick-like. Such as they are
you see them in No. 11, rather clumsy specimens,
while our lady goes on with redoubled energy, now
concentrating her efforts on her front pair of legs

—for when you have six to think of, one pair at a time is about as much as you can easily manage.

In No. 11, the first pair, you will note, is all but free. She wriggles out one of them, and then its fellow. Oh, how she tugs and pulls at them!

NO. 11.—WINGS FREE!

Meanwhile, the tufts of hair on the antennæ, which at first were bunchy and little developed, have begun to expand; she looks, by this time, distinctly more like a respectable insect. Well done, once more; two pairs of legs now free. No. 12 shows them. But, take care; we are getting now rather far out of the mummy-case. Be sure you don't overbalance, and tumble bodily out, tearing your hind pair of legs off, with the force of your fall. Those thin shanks are brittle, and you find little support now from the empty skin and the hollow bracket.

Nature, however, is wiser than her critics. Just

NO. 12.—NOW FOR THE LEGS!

when it looks as if next moment the fly must lose her balance and topple over, she twists suddenly round, with a dexterous lunge, catches the bent stem with two of her free legs, and anchors herself securely. No. 13 shows how this is done. Below is the now almost empty pupa-shell, still enclosing the last two legs, on freeing which our astute little enemy is busily occupied.   But with the two legs on her upper side (as she stands in the illustration) she has caught at the barley-stem, one foot being firmly planted below the bend, and one above it. This gives her a fine purchase to depend upon

NO. 13.—THE LAST PULL ; THE USE OF LEVERAGE.

in her last wild blow for freedom.   A long pull, and a strong pull, and she has got—what the modern woman so ardently craves—complete emancipation !   The third pair of legs are out at last ; she has all the world before her to wander over and lay eggs in.

U

In No. 14 you see her, then, free, but resting. She has now shaken herself out, and left her empty mummy-case imprisoned at her side in the sheath which holds it. Its fate no longer interests her. Then she crawls a little way along the surface of the barley stem, and presently, clasping it with

NO. 14. HANGING HERSELF UP TO DRY.

her four front legs, she hangs herself up, tail downward, to dry in the sunshine. No. 14 graphically represents this curious position. Almost all flying insects, when they emerge from the chrysalis stage, do something analogous. Their wings are still club-like, their antennæ undeveloped or not

fully expanded, their jointed legs weak and groggy.
But after a time, as they breathe or inflate themselves with air, all these parts grow fuller, lighter,
and harder.   The Hessian fly in this predicament
waves her wings to and fro several times across
her back; and in about a quarter of an hour
they have plimmed out fully, so that she can
soar away on her marriage-flight to meet her pro-
spective aërial husband.   As for the tiny silvery
shroud or deserted pupa-case, it is left protruding
from the stem of the barley.

This that I have given you is the history of a
successful and fortunate fly; but not every individual of the species is quite so lucky.   As in the
case of the mosquito, nature at times makes not a
few failures.   Sometimes the flies have insuperable
difficulty in freeing themselves from their articu-
lated coverings; sometimes they break or spoil
their legs or wings, and become helpless cripples.
Yet so strong is the impulse of every species to
fill the world with its like that sometimes, says
Mr. Enock, even these poor maimed insects will
manage to crawl to a proper food-plant, and will
lay their eggs on it bravely like their more fortunate sisters.   He noted one crippled female which
in spite of its feebleness was eighty times over a
happy mother.   This is usually the case with such
small insect pests; their life consists, indeed, of
two things only, eating their way to the winged
stage, and then laying as many eggs as possible, to
do like damage in the next generation.

Three or four hours after emerging, when they

have had time to accustom themselves to the outer air, the male flies soar abroad on gauzy wings to seek their mates; the ladies, on the contrary, are coy, not to say somewhat sluggish, and oftenest wait at home on the under side of a leaf till their lords come to woo them. The well-bred Hessian fly does not gad about to seek a husband. But that is only while she is a maiden; as soon as it comes to laying eggs, she wakes up at once, and takes to business with the utmost energy. She flies off around the fields and looks out a fresh young barley-plant, suitable for a nursery. On its leaves she alights, with her head towards the tip of the blade, and begins depositing her precious burden. When once she has started, she sticks to it for life, using herself up (like our old friend the aphis) in the duties of maternity, and laying as many eggs as she possesses material for. Her conduct, in short, would be exemplary, if she wasted her life on thistles or nettles, and didn't choose to display her maternal affection on the British farmer's barley. So she goes on till she has worn herself out, and often till she has broken three or four of her legs in the pursuit of duty. Then, when she grows quite exhausted, and feels her latter end drawing nigh, she hides herself in the ground—buries herself alive, in fact; and there awaits death with patient resignation.

The average lifetime of the Hessian fly in the adult winged stage seems to be about five days for the females, and probably a good deal less for the males. The bachelors in search of a wife fly sometimes for long distances across country ; but their prospective partners are almost always shyer and more maidenly ; they hide under the leaves and travel but short distances, considering it more lady-like to stop at home and wait for suitors than to go out and seek them. They are not new women. Indeed, so great is their modesty that they often hide in holes in the ground to escape observation ; and they usually alight on the earth, as their colour is blackish, and they are there less exposed to the attacks of birds and other enemies than on the green foliage. It is a noticeable fact in nature that many species of animals seem thus to know in-stinctively the colours with which their own hues will best harmonise, and to poise by preference on such colours ; many dappled or speckled insects, for example, resting with folded wings on the dappled and speckled flower-bunches of the carrot tribe, while green insects affect rather green leaves, and brown or black insects come to anchor on the soil, which best protects them. This is not quite the same thing as what is called protective colour-ing, such as occurs in desert animals, most cf which are spotted like the sand, or in the fishes and crabs which frequent the sargasso-weed in the Sargasso Sea, all of which are of the same pale lemon-yellow tint as the seaweed they lurk among ; for this case of the Hessian fly includes a delibe-rate choice of ingrained habit. The insect has many objects of many different colours spread about in its neighbourhood, but it habitually selects as its resting-place those particular objects which most closely approach its own peculiar ground-tint.

It is a curious fact, however, that in spite of all the apparent pains bestowed upon securing the perpetuation of such destructive creatures as the Hessian fly, the pest itself has its own enemies, as fatal to its life as it is to the barley. Ichneumon flies and other parasites prey by millions on the Hessian fly in its grub condition; and many good authorities believe that the safest way of checking the depredations of the barley-plague is by encouraging the multiplication of its natural enemies. No. 15 shows us one of these industrious little scourges actually at work. She alights on a stem of barley infested by grubs of the Hessian fly, and walks slowly along it, tapping gently as she goes, much as a woodpecker taps with his bill on a tree-trunk to discover the spot where a worm lies buried. After carefully examining the surface, she finds at last a place where something, either in the sound or the feeling of the stem, reveals to her the presence of a Hessian fly grub within the leaf-sheath. Having accurately diagnosed the spot (like a doctor with a stethoscope), she brings her ovipositor (in plain English, her egg-layer) just above the place where the grub is lying snug in its green bed, and pierces the hard leaf-blade with her sharp little lancet. Then she lays her egg in the body of the larva. This egg gives rise in time to a parasitic grub, inside the first one; and the parasite eats out his host's body, and emerges in due time as a full-grown fly, ready to carry on the

same cycle in future. More than nine-tenths of
the Hessian fly grubs hatched out in America
are thus destroyed by parasites before they reach
maturity; and it seems likely that the surest
way of fighting in-
sect plagues like the
Hessian fly is by
encouraging the in-
crease of such natu-
ral destroyers.

At first sight, to
be sure, it may seem
improbable that man
could do anything
to "encourage" the
reproduction of such
very small creatures ;
but that is not really
so. All that is ne-
cessary is to keep
the straw in which
the parasitic grubs
abound, and so allow
the two hostile kinds
to fight it out among
themselves for the
farmer's benefit.
Mr. Enock mentions

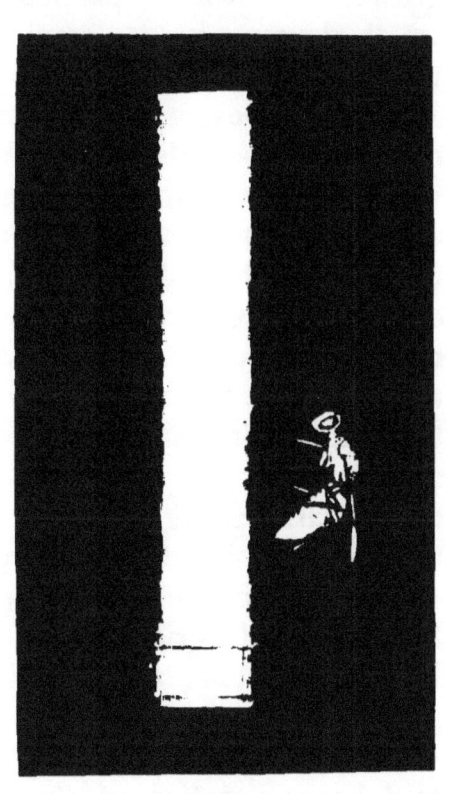

NO. 15.—WILY ENEMY LAYING HER EGGS
IN THE LARVA.

an instructive case of this sort from America,
where the Californian orange-growers were
almost being ruined by the depredations of
the scale-insect, a queer little beast which you

may often find on the rind of certain imported oranges. But an enemy to the scale-insect was discovered in Australia—an enemy to the scale-insect, and therefore an ally of the harassed orange-grower. It was a particular kind of ladybird, which devours in its larval stage whole tribes of the scale-insects. That wonderful entomologist, Professor Riley, whose services were worth many millions of pounds to the American farmers, got wind betimes of this new destroyer, and imported a few specimens, actually sending a skilled agent to Australia to collect them. The precious little creatures were housed at once in a muslin tent, covering a scale-infested orange tree; and there, rising to a sense of the duty imposed upon them, they laid their eggs on the leaves with commendable promptitude. The larvæ soon hatched out, and began feeding upon the scale-insects; and in an incredibly short time there were beetles enough on that single tree to distribute by boxfuls among the distressed agriculturists. The result was that before very long the scale-insect became a rare specimen in California. But that was in the United States; English folk are too "practical" to take any notice of those theoretical men of science. They put their hands in their pockets and let their crops get destroyed in the good old "practical" way; then they shake their heads and observe with a smile that "there are great diffi-culties" in the way of doing anything.

PUBLICATIONS OF
GEORGE NEWNES
LIMITED

7-12 SOUTHAMPTON ST.
STRAND, LONDON, W.C.

11/98

Illustrated Books.

FARTHEST NORTH.

Being the Record of a Voyage of Exploration of the ship *Fram*, 1893-96, and of a Fifteen Months' Sledge Journey by Dr Nansen and Lieut. Johansen. With an Appendix by Otto Sverdrup, Captain of the *Fram*. Popular edition. In 2 vols., royal 8vo. With about 120 full-page and 90 text Illustrations. Coloured Plate and Map, cloth extra, 17s.

"The narrative of one of the most remarkable and adventurous voyages of discovery that have been made."—*Scotsman*.

"A masterpiece of story-telling."—*Times*.

FLASHLIGHTS ON NATURE.

By GRANT ALLEN. With 150 Illustrations by FREDERICK ENOCK. Crown 8vo, cloth extra, 6s.

"Charming and romantic scientific chats."—*Review of Reviews*.

DOWN THE STREAM OF CIVILIZATION.

By WORDSWORTH DONISTHORPE. With 104 Illustrations from Photographs. Crown 8vo, cloth extra, 6s.

WELLINGTON AND WATERLOO.

By Major ARTHUR GRIFFITHS. With an Introduction by Field-Marshal Viscount WOLSELEY. Profusely Illustrated. Demy 4to, cloth extra, 10s. 6d.

"Pen and camera have worked well together in this handsome quarto volume."—*Daily News*.

"An important volume, excellently calculated to make a charming gift-book."—*St James's Gazette*.

RAIDERS AND REBELS IN SOUTH AFRICA.

By Mrs GOODWIN GREEN. With 14 full-page Illustrations by the Author. Crown 8vo, cloth extra, 5s.

ENGLAND'S HISTORY, AS PICTURED BY FAMOUS PAINTERS.

An Album of 256 Historical Pictures, with descriptive text. Edited by A. G. TEMPLE, F.S.A. Oblong 4to, cloth extra, gilt leaves, 10s. 6d.

". . . Large photographic reproductions of more or less notable pictures, chronologically arranged, commencing with Lord Leighton's 'Commerce Between the Ancient Britons and the Phœnicians,' and ending with Mr C. W. Furse's spirited picture of 'The Battle of Doornkop—from the Boer Position,' thus presenting, from the earliest times down to yesterday, notable scenes in our history as realised by artists of good standing."—*Birmingham Daily Post*.

ENGLISH CATHEDRALS ILLUSTRATED.

By FRANCIS BOND, M.A., F.G.S., Hon. A.R.I.B.A. With 188 Illustrations from Photographs. Crown 8vo, cloth extra. *[In the press.*

LONDON : GEORGE NEWNES LIMITED, PUBLISHERS.

Illustrated Books.

THE ART BIBLE.

Comprising the text of the Old and New Testaments, printed in entirely new type, specially selected for its clearness and sharpness of outline, and with 850 Illustrations, Maps, &c.

In One Volume, 1,360 pp. super royal 8vo, handsomely bound in cloth, gilt leaves, 12s. ; French morocco, bevelled boards, tooled in gold and blind, gilt leaves, 16s.; plain Persian morocco, gilt leaves, 18s. ; Persian morocco, antique scroll in gold, gilt leaves, 21s. ; limp morocco, Yapp style, flexible back, solid gold leaves, 30s. ; Turkey morocco, hand-tooled in gold, solid gold leaves, 38s.

OLD TESTAMENT, with 660 Illustrations, handsomely bound in cloth, gilt leaves, 9s.

NEW TESTAMENT, with 190 Illustrations, handsomely bound in cloth, gilt leaves, 5s.

CANON BASIL WILBERFORCE writes:—"I think that the illustrations are excellent, and that in producing this book at such a low price you have done a public service, and materially added to the interest of the study of the Scriptures."

THE SWISS FAMILY ROBINSON.

A New Version by E. A. BRAYLEY HODGETTS, with 100 Illustrations by J. FINNEMORE. Royal 8vo, cloth extra, gilt leaves, 10s. 6d.

"Will cause delight in many households."—*Westminster Gazette.*

"By far the best and most handsome translation that has yet appeared . . . taken direct from the German and the language is far more terse and effective than that of those from the French version."—*Standard.*

ROUND THE WORLD, FROM LONDON BRIDGE TO CHARING CROSS, VIA YOKOHAMA AND CHICAGO.

An Album of 284 Pictures from Photographs of the Chief Places of Interest in all Parts of the World, with descriptive text. Oblong 4to, cloth extra, gilt leaves, 10s. 6d.

"We are taken through all the principal cities and picturesque and historic places from Calais to Brindisi, we mentally travel up the Nile, we go through all the great commercial and historic scenes in India, Malaysia, China, and Japan before crossing to the American Continent, we are given views of the forests, rivers, and mountains of the new world, the cities that are dotted over its bosom, and all the sights from the Falls of Niagara down to the silver mountain of Potosi and the streets of Buenos Ayres. In short, the work is as complete as it is possible to make it."—*The Stock Exchange.*

LONDON : GEORGE NEWNES LIMITED, PUBLISHERS.

Illustrated Books.

ROUND THE COAST.

An Album of 284 Pictures from Recent Photographs of the Watering Places and Resorts in the United Kingdom, with descriptive text. Oblong 4to, cloth extra, gilt leaves, 10s. 6d.

"We know nothing at anything like the price that can be compared with these for giving to the sedentary traveller veracious glimpses of what the world or his own seashores contains that is interesting and picturesque."—*Times*.

ROUND LONDON.

An Album of 284 Pictures from Photographs of the Chief Places of Interest in and around London, with descriptive text. Oblong 4to, cloth extra, gilt, 10s. 6d.

"The illustrations are taken from photographs, of the most notable and characteristic of the metropolitan sights. Few quarters of London or aspects of London life are neglected; its business and its pleasures, its architecture and its street traffic receive illustration in all their phases. The photographs have been admirably reproduced."—*Scotsman*.

THE THAMES ILLUSTRATED. A Picturesque Journeying from Richmond to Oxford. 164 large and 170 small Photographic Plates, with descriptive text, 4to, cloth extra, gilt leaves, 10s. 6d.

". . . A marvel at the price. The illustrations are very numerous and exceedingly beautiful, every picturesque nook and cranny having been sought out, and no such charming souvenir of the Thames or incitement to enjoy its beauties is in existence."—*Army and Navy Gazette*.

ALL ABOUT ANIMALS.

260 Illustrations of Animal Life, from Photographs by Gambier Bolton, F.Z.S., and others, with explanatory text. Oblong 4to, cloth extra, gilt leaves, 10s. 6d.

"Not a dry and scientific work, but brightly written and well illustrated. Well illustrated it certainly is. Mr Gambier Bolton's photographs of lions, tigers, leopards, and other feline creatures have never been surpassed, and hundreds of such pictures are to follow. It is really a wonderful production."—*Army and Navy Gazette*.

ZIG-ZAGS AT THE ZOO.

By Arthur Morrison and J. A. Shepherd. 236 pp., super royal 8vo, cloth extra, 7s. 6d.

"'Zig-Zags at the Zoo' is a necessary volume to all people who are fond of animals and gifted with a sense of humour."—*Spectator*.

"A most delightful book."—*Glasgow Herald*.

"Charming volume."—*World*.

"Letterpress and illustrations are both replete with drollery and harmless fun."—*Morning Post*.

LONDON: GEORGE NEWNES LIMITED, PUBLISHERS.

Illustrated Books.

NEW GROUND IN NORWAY.

Ringerike — Telemarken — Sætersdalen. By E. J. GOODMAN. With numerous Illustrations. Demy 8vo, cloth extra, 10s. 6d.

" Full of information as to the less frequented parts of Southern Norway. . . . Well printed and capitally illustrated."—*Pall Mall Gazette.*

THE WAY OF THE CROSS: A PICTORIAL PILGRIMAGE FROM BETHLEHEM TO CALVARY.

240 Views of the Holy Land from Photographs, with descriptive text. Oblong 4to, cloth extra, gilt leaves, 8s. 6d.

" Some remarkably fine views."—*Standard.*
" This triumph of art, whose every leaf preaches a sermon and brings us in touch with the past."—*Catholic Times.*

PRETTY HOMES.

By Mrs HORSFALL. 168 pp. With 60 Illustrations. 8vo, cloth, 3s. 6d.

THE STRAND MAGAZINE.

An Illustrated Magazine. Edited by GEORGE NEWNES. Vols. 12 to 15, cloth extra, gilt leaves. Price 6s. 6d. each. [Vols. 1 to 11 are out of print.]

WIDE WORLD MAGAZINE.

Vol. 1., cloth extra, gilt leaves, 6s. 6d.

THE NAVY AND ARMY ILLUSTRATED.

Vols. 1 to 5, bound in cloth extra, gilt leaves, 12s. each. Vol. 6, 18s.

COUNTRY LIFE ILLUSTRATED.

The Journal for all interested in Country Life and Country Pursuits. Profusely Illustrated. Vols. 1 to 3, folio cloth, gilt leaves, 21s. ; half morocco, 25s.

THE HOME MAGAZINE.

Vol. 1, cloth, 5s.

LONDON : GEORGE NEWNES LIMITED, PUBLISHERS.

The New Atlas.

THE CITIZEN'S ATLAS.

Comprising 100 Maps and Gazetteer. Edited by J. G. BARTHOLOMEW, F.R.G.S. Crown folio, cloth extra, 16s. net ; half-morocco, 18s. 6d. net.

The Stratford-on-Avon Shakespeare.

THE WORKS OF WILLIAM SHAKESPEARE.

With Glossarial Side-notes. Complete in 12 vols., bound in cloth, with cut or uncut edges, 18s. ; or, enclosed in a quaint cloth box, 21s. ; also in half-morocco, gilt top, 55s. ; crushed grained Persian morocco, in box, 70s., or straight paste grained, gilt tops, in box, with steel clasp, 50s.

"We can unhesitatingly say that the Newnes edition is quite the most handy and readable edition which we have as yet seen—and the number of editions with which we are acquainted is legion. A wise discretion has been exercised as to the size of the volume, which is not too small—as is the case with many existing 'pocket' editions—whilst the semi-antique 'laid' paper, and the absolutely perfect typography, reflect great credit both on the firm which initiated the work and on the printers who produced it. It is not encumbered with notes, but all needful glossarial explanations are ingeniously embedded in small type in the text."—*Broad Arrow*.

"A handy, well printed edition."—*Athenæum*.

"Admirable little edition."—*Star*.

"The typography is excellent."—*Globe*.

"There could be no better edition for general use."—*Scotsman*.

"The size is suitable alike for the bookshelf and the pocket. The type of the text is bold and clear, on antique paper. The tiny side-notes are explanatory only of such obscure words and phrases as are not readily explained by the context. The title-page, which follows the lines of that of the first folio edition, may be taken as an indication of an intention to follow as closely as possible the text of the early editions, which were 'the freshest from Shakespeare's own hands.'"—*Daily News*.

LONDON : GEORGE NEWNES LIMITED, PUBLISHERS.

Specimen Page of the

Stratford-on-Avon   Shakespeare
Hamlet

12 Volumes strongly bound in cloth, 18s.

12 volumes, in cloth box, for Presents, 21s.

Ham. The king doth wake to-night and takes
    his rouse,
Keeps wassail, and the swaggering upspring
    reels ;
And as he drains his draughts of Rhenish down,
The kettle-drum and trumpet thus bray out
The triumph of his pledge.
    Hor.   Is it a custom?
    Ham. Ay, marry is't ;
But to my mind, though I am native here
And to the manner born, it is a custom
More honour'd in the breach than the observ-
    ance.
This heavy-headed revel east and west   [blamed
Makes us traduc'd and tax'd* of other nations :
They clepe* us drunkards, and with swinish [call
    phrase
Soil our addition*; and indeed it takes   [title
From our achievements, though perform'd at
    height,
The pith and marrow of our attribute.
So, oft it chances in particular men,
That for some vicious mole of nature in them,
As, in their birth—wherein they are not guilty,
Since nature cannot choose his origin—
By the o'ergrowth of some complexion,*
                        [natural disposition
Oft breaking down the pales and forts of reason,
Or by some habit that too much o'er-leavens
The form of plausive manners, that these men,
Carrying, I say, the stamp of one defect,
Being nature's livery, or fortune's star,—
Their virtues else—be they as pure as grace,
As infinite as man may undergo*—   [experience
Shall in the general censure take corruption
From that particular fault: the dram of eale*[? evil

LONDON : GEORGE NEWNES LIMITED, PUBLISHERS.

The
Library of Useful Stories.

Small 8vo, cloth, price 1s. each Volume, post free 1s. 2d.

" *The more Science advances, the more it becomes concentrated in little books.*"—LEIBNITZ.

I.
THE STORY OF THE STARS.

By G. F. CHAMBERS, F.R.A.S., Author of " Handbook of Descriptive and Practical Astronomy," &c. With 24 Illustrations.

" Mr Chambers writes in a vigorous and attractive style, and shows himself able to combine to an uncommon degree scientific accuracy of statement with a clear and attractive exposition. Beginners in astronomy who wish to acquaint themselves merely with the outlines of a noble science will find this volume of real service."—*Speaker.*

" Told in a pleasing and attractive manner."—*Athenæum.*

II.
THE STORY OF PRIMITIVE MAN.

By EDWARD CLODD, Author of " The Story of Creation," &c. With 88 Illustrations.

" It possesses the chief qualities that go to make a good book for the average man."—*Nature.*

" Well printed, well bound, profusely illustrated, and in every respect capital material, on one of the most progressive of sciences."—*Daily Chronicle.*

III.
THE STORY OF THE PLANTS.

By GRANT ALLEN. With 49 Illustrations.

" A brightly written, clear and accurate summary of the functions and habits of plants."—*Daily Chronicle.*

" The whole book is excellent, but special praise is due to his exposition of the relations existing between plants and insects. Many chapters of the story he tells must prove to the uninitiated as exciting as a romance."—*Aberdeen Free Press.*

IV.
THE STORY OF THE EARTH IN PAST AGES.

By H. G. SEELEY, F.R.S., Professor of Geology, Geography, and Mineralogy in King's College, London. With 40 Illustrations.

" A simple and popular summing up of the results that have been reached by geological science."—*Scotsman.*

" Told plainly and pleasantly for a popular audience."—*Bookman.*

LONDON : GEORGE NEWNES LIMITED, PUBLISHERS.

The Library of Useful Stories.

V.
THE STORY OF THE SOLAR SYSTEM.
By G. F. CHAMBERS, F.R.A.S. With 28 Illustrations.

"His descriptions possess the double quality of simplicity and attractiveness."—*Nature.*

"He repudiates the idea that unless a man can command a big telescope he is not in a position to do useful work in astronomy. . . . The little volume is an admirable example of science made easy without the sacrifice of strict accuracy of statement."—*Speaker.*

VI.
THE STORY OF A PIECE OF COAL.
By E. A. MARTIN. With 38 Illustrations.

"Treated with wonderful skill, simplicity, and thoroughness."—*Bookseller.*

"Explains in simple and delightful fashion what coal is, whence it comes, and whither it goes, and in the concluding chapters shows how intimately it is connected with the interests of the botanist, the geologist, the physicist, the chemist, and the merchant."—*Bradford Observer.*

VII.
THE STORY OF ELECTRICITY.
By J. MUNRO, Joint Author of "The Pocket-book of Electrical Rules and Tables." With 100 Illustrations.

"Just the kind of book to give the general reader more correct views of the subject than many a pretentious tome."—*The Electrical Engineer.*

"For general interest we must pronounce the little book without a peer, style and matter being alike excellent."—*Glasgow Daily Mail.*

"A handy little book which has certainly the great merit of being up to date. We anticipate a large demand for the book."—*Electricity.*

VIII.
THE STORY OF EXTINCT CIVILIZATIONS OF THE EAST.
By R. E. ANDERSON, M.A., contributor to Chambers' Encyclopædia, Encyclopædia Britannica, and Dictionary of National Biography, &c. With Maps.

"The author has performed a much needed service in a masterly manner. . . . We have nothing but praise for the work."—*Literary World.*

"An admirable compendium of a department of knowledge which has been greatly advanced by the research of recent years."—*Aberdeen Free Press.*

LONDON: GEORGE NEWNES LIMITED, PUBLISHERS.

The Library of Useful Stories.

IX.
THE STORY OF THE CHEMICAL ELEMENTS.
By M. M. Pattison Muir, M.A., Fellow and Prælector in Chemistry of Gonville and Caius College, Cambridge.

" One of the most perfect popular introductions to science extant."
—*British Medical Journal.*

" Prof. Muir tells an enthralling story of the wonderful transformations of matter under the chemist's magic wand. Ignoring formulæ he appeals in homely phrase to the imagination of the reader."—*Knowledge.*

X.
THE STORY OF FOREST AND STREAM.
By James Rodway, F.L.S., Author of " In the Guiana Forest," &c. With 27 Illustrations.

" Contains a short description of a tropical forest, together with some elementary lessons which can be learned by studying the incessant struggle for existence of its varied flora."—*Academy.*

" A noteworthy addition to the series in which it appears."—*Scotsman.*

XI.
THE STORY OF THE WEATHER.
By G. F. Chambers, F.R.A.S., of the Inner Temple, Author of "Story of the Stars," &c. With 50 Illustrations.

" An interesting volume about weather, and especially English weather, and presents facts, ideas, and suggestions which ordinary people will be glad to know."—*St James's Budget.*

" Shows how the weather forecasts are drawn up at the Meteorological Office, explains the construction and use of the various meteorological instruments, describes the nature and causes of such phenomena as the aurora borealis, and gives a collection of weather facts and signs."—*Literary World.*

XII.
THE STORY OF THE EARTH'S ATMOSPHERE.
By Douglas Archibald, M.A., Fellow and sometime Vice-President of the Royal Meteorological Society, London. With 44 Illustrations.

" One of the best of the Story series that we have read . . . the author is frequently able from his wide travels to illustrate his remarks from his own personal experience in climates where meteorological manifestations can be witnessed on a grander scale than in our own country."—*Nature.*

LONDON : GEORGE NEWNES LIMITED, PUBLISHERS.

The Library of Useful Stories.

LONDON: GEORGE NEWNES LIMITED, PUBLISHERS.

The
Library of Useful Stories.

XVI.
THE STORY OF LIFE IN THE SEAS.
By SYDNEY J. HICKSON, D.Sc., F.R.S., Professor of Zoology in the Owen's College, Manchester. 42 Illustns.

" Prof. Hickson is assuredly to be congratulated on the very able manner in which he has acquitted himself of his task. In the course of eight chapters, written in pleasant and unaffected style, he gives a fairly complete sketch of the vertebrate and invertebrate fauna associated with different marine regions, the surface, the shallow water, and the depths."—*Saturday Review.*

"Such books as these lay the reader under a deep obligation to writers of Dr Hickson's eminence in the scientific world."—*Spectator.*

XVII.
THE STORY OF PHOTOGRAPHY.
By A. T. STORY. With Illustrations.

"An interesting popular sketch of a subject the more common books about which are generally of a purely technical character, designed to inform amateurs. Mr Story does not profess to give instructions in the art ; but his history of its progress, his description of the various forms of apparatus and the various processes, and his statements of the relation between photography and the finer arts cannot but impart an intelligent interest in this versatile handmaiden of science."—*Scotsman.*

XVIII.
THE STORY OF RELIGIONS.
By E. D. PRICE, F.G.S.

XIX.
THE STORY OF THE COTTON PLANT.
By F. WILKINSON, F.G.S., Director of the Textile and Engineering School, Bolton. With 38 Illustrations.

XX.
THE STORY OF GEOGRAPHICAL DISCOVERY.
By JOSEPH JACOBS. With 24 Maps, etc.

XXI.
THE STORY OF THE MIND.
By Professor J. M. BALDWIN.

XXII.
THE STORY OF THE BRITISH RACE.
By JOHN MUNRO. With four Maps.

LONDON : GEORGE NEWNES LIMITED, PUBLISHERS.

Conan Doyle's Stories.

THE EXPLOITS OF BRIGADIER GERARD.

With 24 Illustrations by W. B. Wollen. Crown 8vo, cloth, 6s.

" In these days of pessimistic problem novels, when the element of romance seems to be fading out of fiction, it is delightful to come upon these tales and glories of a soldier's life. They are buoyant, vital, steeped in the stir and freshness of the open air, abounding in tragedy and gaiety. . . . It is a fascinating book, and one to be read."—*Daily News.*

ADVENTURES OF SHERLOCK HOLMES.

With 25 Illustrations by Sidney Paget. Crown 8vo, cloth, 3s. 6d.

" For those to whom the good, honest, breathless detective story is dear, Dr Doyle's book will prove a veritable godsend."—*Athenæum.*

LAST ADVENTURES OF SHERLOCK HOLMES.

With 25 Illustrations by Sidney Paget. Crown 8vo, cloth, 3s. 6d.

" Should become a favourite gift book."—*Liverpool Mercury.*

THE SIGN OF FOUR.

An Earlier Adventure of Sherlock Holmes. Crown 8vo, cloth, 3s. 6d.

" The 'Adventures of Sherlock Holmes' should be read by all who desire to improve their faculty of observation. Fathers would do well to make it a birthday present to their boys, and if they do this, they certainly may have the comforting thought that the book will be read from beginning to end."—*Glamorgan Gazette.*

THE ROMANCE OF HISTORY.

By HERBERT GREENHOUGH SMITH. 292 pp., crown 8vo, cloth, 3s. 6d.

A series of graphic sketches of the leading incidents in the lives of Masaniello, Prince Rupert, Marino Faliero, Bayard, Lithgow, Jacqueline de Laguette, Vidocq, Lochiel, Casanova. The volume is printed on antique paper, and bound in old style with uncut edges.

" Seldom has the old adage that truth is stranger than fiction been better exemplified than in some of the incidents related in ' The Romance of History.' The book is well written and both the subjects selected and the way in which they are treated leave little to be desired."—*Morning Post.*

" Pre-eminently interesting, bright, clear and attractive."—*Daily Chronicle.*

LONDON : GEORGE NEWNES LIMITED, PUBLISHERS.

Popular Novels.

AT THE SIGN OF THE GOLDEN HORN.
By JOHN K. LEYS. Crown 8vo, cloth, 3s. 6d.

SHAFTS FROM AN EASTERN QUIVER.
By CHARLES J. MANSFORD. With 25 Illustrations by
Alfred Pearse. Cloth, 3s. 6d.

" Mr Mansford has the gift of a story-teller, and he uniformly
writes like a scholar. . . . The illustrations, though small, are admirably executed, and enhance the piquancy—though that was hardly
needed—of the letterpress."—*Spectator*.

THE BEECHCOURT MYSTERY.
By CARLTON STRANGE. Cloth, 3s. 6d.

" A novel and well-constructed plot."—*Liverpool Courier*.

WHAT'S BRED IN THE BONE.
By GRANT ALLEN. Cloth, 3s. 6d.

HEARTS OF GOLD AND HEARTS OF STEEL.
By the late HENRY HERMAN. Cloth, 3s. 6d.

FOR GOD AND THE CZAR.
A Story of Jewish Persecutions in Russia. By J. E.
MUDDOCK. Cloth, 3s. 6d.

ONLY A WOMAN'S HEART.
The Story of a Woman's Love : A Woman's Sorrow.
By J. E. MUDDOCK. Cloth, 3s. 6d.

" Has an air of heartiness about it, and its plot is well worked out."
—*Academy*.

**THE RUBIES OF RAJMAR ; or, MR CHARLECOTE'S
DAUGHTERS.**
A Romance. By Mrs EGERTON EASTWICK (Pleydell
North). Cloth, 3s. 6d.

" Throughout, the plot is well conceived, its treatment is terse and
vigorous, and the series of exciting incidents by which the *dénouement*
is reached, form a narrative well worth reading."—W. LE QUEUX in
The Literary World.

LONDON : GEORGE NEWNES LIMITED, PUBLISHERS.

Popular Novels.

THE KING OF THE BRONCOS, AND OTHER STORIES OF NEW MEXICO.

By CHAS. J. LUMMIS. With Portrait and 8 full-page Illustrations. Cloth, 5s.

"The author has fallen under the spell of the wilderness, and writes of it with affection. . . . All the stories are excellent reading, and some of them are dramatic."—*Manchester Guardian.*

"All boys will enjoy 'The King of the Broncos' . . . truly exciting stories of a West which is still largely deserving of the title of wild."—*Daily Telegraph.*

THE KING'S OAK, AND OTHER STORIES.

By ROBERT CROMIE, Author of "The Crack of Doom," &c. 130 pp., 1s. Cloth, 2s.

"Five well-written and entertaining stories."—*Literary World.*
"Short stories, bright and dramatic."—*To-day.*
"A capital collection of short stories."—*Black and White.*

THE LOST LINER.

By ROBERT CROMIE. Crown 8vo, cloth, 3s. 6d.

STORIES FROM THE DIARY OF A DOCTOR.

By L. T. MEADE and CLIFFORD HALIFAX, M.D., Authors of "The Medicine Lady." With 24 Illustrations by A. PEARSE. Cloth extra, 6s.

"Cleverly-planned and brightly-told stories."—*Bradford Observer.*
"They are well told and salient in every feature."—*Leeds Mercury.*

MEMOIRS OF A MOTHER-IN-LAW.

By GEORGE R. SIMS. Cloth, 2s. 6d.

"This is a pleasant sample of 'Dagonet's' semi-humorous writings. He has a peculiar talent of finding amusement in experiences relating to dwellings, servants, shopkeepers, tradespeople, and other folk connected with the domestic household, and the 'Mother-in-Law' in his new book deals in a very masterful way with all the foregoing subjects, and many more besides."—*Freeman's Journal.*

TWO GIRLS.

By AMY E. BLANCHARD. With Illustrations by Ida Waugh. Cloth extra, 3s. 6d.

"A delightful addition to the girls' bookshelf."—*Gentlewoman.*
"A bright sparkling story for girls, brimful of innocent fun."—*Liverpool Courier.*

LONDON: GEORGE NEWNES LIMITED, PUBLISHERS.

The Queen and her Reign.

PIONEER WOMEN IN VICTORIA'S REIGN.

Being short Histories of Great Movements. By EDWIN A. PRATT. Crown 8vo, cloth, 5s.

"A survey given with great skill and effect."—*Times.*
"His chapters on Women's Work in Emigration and in Medicine are admirable."—*Pall Mall Gazette.*
"The nursing record of Queen Victoria's Reign, ably told here, will interest so many people just now."—*St James's Gazette.*

A WOMAN'S WORK FOR WOMEN.

Being an account of the Philanthropic Work of Miss L. M. HUBBARD. By E. A. PRATT. Small crown 8vo, cloth.

QUEEN VICTORIA'S DOLLS.

By FRANCES H. LOW. With 40 Full-page Coloured Illustrations and numerous Sketches and Initial Letters, by ALAN WRIGHT. Cheap Edition, crown 4to, 5s.

"No one who has not perused this entertaining record can in reality appreciate the diligent, alert child-life of Britain's truest gentlewoman. The full-page coloured illustration, showing the dolls in their gorgeous costumes, and wooden attitudes, are almost as naïve as they are excellent."—*The Gentlewoman.*

THE PRINCESS OF WALES: A BIOGRAPHICAL SKETCH.

By MARY SPENCER-WARREN. With Portraits of the Princess at various periods, and Illustrations from Photographs taken in Denmark, and at Sandringham, Marlborough House, &c. With 53 Portraits and Illustrations. Crown 8vo, cloth extra, 5s.

"An excellent biography . . . narrated with admirable simplicity and lucidity."—*Westminster Review.*

HEROES OF THE VICTORIA CROSS.

By T. E. TOOMEY, late Colour-Sergeant. A record of the "Cross" and its Wearers, with Narratives of Daring Deeds, and 228 Portraits. Crown 8vo, cloth, 5s.

"The value, utility, and interest of the book are obvious."
—*Liverpool Courier.*

THE VICTORIAN ERA: A GRAPHIC RECORD OF A GLORIOUS REIGN.

By R. E. ANDERSON, M.A. With 136 Illustrations and with Photographic Portrait of the Queen. Cloth, 2s.

LONDON : GEORGE NEWNES LIMITED, PUBLISHERS.

THE ORACLE ENCYCLOPÆDIA.
Profusely Illustrated. Edited by R. W. EGERTON
EASTWICK, B.A. (of the Middle Temple), complete in
5 vols., price 30s., or in half morocco, 52s. 6d.

SONGS OF CHILDHOOD.
Verses by EUGENE FIELD. Music by REGINALD
DE KOVEN and others. Music 4to, cloth extra, gilt
leaves, 7s. 6d.

"It is the best compliment to the music to say that its pretty tune-
fulness is not unworthy of the words. There are songs here that will
go straight to a child's heart, and not one that will miss the child-
lover."—*Pall Mall Gazette*.

A LITTLE BOOK OF PLAYS, FOR PROFES-
SIONAL AND AMATEUR ACTORS.
Adapted from the French by CONSTANCE BEERBOHM.
With 22 Illustrations of Scenes. Paper covers, 1s.

"An excellent collection of short dramatic sketches, suited to the
requirements of amateurs."—*Black and White*.

AIDS TO HEALTH AND BEAUTY.
A Complete Toilet Guide. By MIRANDA. Long
8vo, 1s.

WRINKLES FOR CYCLISTS.
By G. LACY HILLIER. Small crown 8vo, 1s.

THE HUB CYCLING MAP OF ENGLAND AND
WALES.
By J. BARTHOLOMEW, F.R.G.S. Printed in colours
and folded in pocket case, 6d.; mounted on linen, 1s.

THE COAST TRIPS OF GREAT BRITAIN.
Prefaced by a Description of the Thames Scenery
from London Bridge to the Nore. Compiled by
MILTON SMITH, and profusely illustrated with Original
Sketches by W. T. WHITEHEAD. 6d. net.

6,000 TIT-BITS OF CURIOUS INFORMATION.
Being 6,000 Answers to 6,000 Questions from the
Enquiry Column of *Tit-Bits*, in 6 vols., price 2s. 6d.
each. [Vol. 1 out of print.]
‗ Each Volume complete in itself.

LONDON : GEORGE NEWNES LIMITED, PUBLISHERS.

Serials now in course of Publication.

THE STRAND MAGAZINE.
6d. monthly ; cases for binding, 1s.

WIDE WORLD MAGAZINE.
6d. monthly ; cases for binding, 1s. 6d.

TIT-BITS. 1d. weekly ; cases for binding, 1s.

HOME MAGAZINE.
A Journal for Sunday and week-day reading; 1d. weekly ; cases for binding, 1s. 6d.

WOMAN'S LIFE. 1d. weekly ; cases for binding, 1s.

THE HUB.
1d., an Illustrated Weekly Journal for Wheelmen and Women ; cases for binding, 1s.

THE NAVY AND ARMY ILLUSTRATED.
6d. weekly ; cases for binding, 2s. 6d.

COUNTRY LIFE ILLUSTRATED.
6d. weekly ; cases for binding, 2s. 6d., cloth ; 6s., half morocco.

LADIES' FIELD. 6d. weekly.

THE CITIZEN'S ATLAS AND GAZETTEER.
6d. fortnightly ; cases for binding, 3s., cloth ; 4s. 6d., half morocco.

HANS ANDERSEN'S FAIRY TALES.
With upwards of 400 Illustrations by HELEN STRATTON. In fourteen fortnightly parts, price 7d. each.

THE PENNY LIBRARY OF FAMOUS BOOKS.
1d. Weekly.

Notable Books.

I.

THE LAND OF THE MIDNIGHT SUN.
With Map and 250 Illustrations. In 12 Fortnightly Parts, 6d. each.

LONDON : GEORGE NEWNES LIMITED, PUBLISHERS.

Sixpenny Editions of Copyright Novels.

Well printed, on good paper, 8vo, sewed.

THE ADVENTURES OF SHERLOCK HOLMES.
By A. CONAN DOYLE.

LAST ADVENTURES OF SHERLOCK HOLMES.
By A. CONAN DOYLE.

THE SIGN OF FOUR: AN EARLY ADVENTURE
OF SHERLOCK HOLMES.
By A. CONAN DOYLE.

ROBERT ELSMERE.
By Mrs HUMPHREY WARD.

LONDON : GEORGE NEWNES LIMITED, PUBLISHERS.

FISH

DOGS

HORSES

TENNIS

GOLF

RACING

SOCIETY

HUNTING

AND ALL SORTS OF

OUT-DOOR SPORT

ARE FEATURES OF

COUNTRY LIFE

The Most Beautifully

Illustrated Paper in

the World

Price Sixpence

LONDON: GEORGE NEWNES LIMITED, PUBLISHERS.